Quality-of-Service Aware Design and Management of Embedded Mixed-Criticality Systems

Behnaz Ranjbar • Alireza Ejlali • Akash Kumar

Quality-of-Service Aware Design and Management of Embedded Mixed-Criticality Systems

Springer

Behnaz Ranjbar
TU Dresden
Dresden, Germany

Alireza Ejlali
Sharif University of Technology
Tehran, Iran

Akash Kumar
TU Dresden
Dresden, Germany

ISBN 978-3-031-38962-7 ISBN 978-3-031-38960-3 (eBook)
https://doi.org/10.1007/978-3-031-38960-3

This Springer imprint is published by the registered company Springer Nature Switzerland AG
The registered company address is: Gewerbestrasse 11, 6330 Cham, Switzerland

Paper in this product is recyclable.

Dedicated to our families

Summary

Quality-of-Service Aware Design and Management of Embedded Mixed-Criticality Systems

Nowadays, implementing a complex system, which executes various applications with different levels of assurance, is a growing trend in modern embedded real-time systems to meet cost, space, timing, and power consumption requirements. Medical devices, automotive, and avionics industries are the most common safety-critical applications, exploiting these systems known as Mixed-Criticality (MC) systems. MC applications are real time, and to ensure the correctness of these applications, it is essential to meet strict timing requirements as well as functional specifications. The correct design of such MC systems requires a thorough understanding of the system's functions and their importance to the system. A failure/deadline miss in functions with various criticality levels has a different impact on the system, from no effect to catastrophic consequences. Failure in the execution of tasks with higher criticality levels (HC tasks) may lead to system failure and cause irreparable damage to the system, while although Low-Criticality (LC) tasks assist the system in carrying out its mission successfully, their failure has less impact on the system's functionality and does not harm the system itself to fail.

In order to guarantee the MC system safety, tasks are analyzed with different assumptions to obtain different Worst-Case Execution Times (WCETs) corresponding to the multiple criticality levels and the operation mode of the system (e.g., low WCET and high WCET). If the execution time of at least one HC task exceeds its low WCET, the system switches from low-criticality mode (LO mode) to high-criticality mode (HI mode). Then, all HC tasks continue executing by considering the high WCET to guarantee the system's safety. In this HI mode, all or some LC tasks are dropped/degraded in favor of HC tasks to ensure HC tasks' correct execution. Here, if we consider very low values for low WCETs, more LC tasks are guaranteed to be scheduled in a processor at design-time. However, it may cause frequent mode switches and drop more LC tasks at run-time due to inefficient low WCET determination. On the other hand, by using a larger low WCET, fewer LC tasks are scheduled in the LO mode, which under-utilizes the processor. To this end, determining an appropriate low WCET for each HC task is crucial in designing efficient MC systems and ensuring Quality-of-Service (QoS) maximization (i.e., execute more LC tasks). However, in the case where the low WCETs are set correctly, it is not recommended to drop/degrade the LC tasks in the HI mode due to its negative impact on the other functions or on the entire system in accomplishing its mission correctly. Therefore, how to model the MC tasks and analyze the task dropping in the HI mode are significant challenges in designing efficient MC systems that must be considered to guarantee the successful execution of all HC tasks to prevent catastrophic damages while improving the QoS.

Due to the continuous rise in computational demand for MC tasks in safety-critical applications, like controlling autonomous driving, the designers are motivated to deploy MC applications on multi-core platforms. Although the parallel execution feature of multi-core platforms helps to improve QoS and ensures the real-timeliness, high power consumption and temperature of cores may make the system more susceptible to failures and instability, which is not desirable in MC applications. Therefore, improving the MC system's QoS while managing the power consumption and guaranteeing real-time constraints is the critical issue in designing such MC systems in multi-core platforms.

This book addresses the mentioned challenges associated with efficient MC system design. We first focus on application analysis by determining the appropriate WCET by proposing two novel approaches to provide a reasonable trade-off between the number of scheduled LC tasks at design-time and the probability of mode switching at run-time to improve the system utilization and QoS. The first approach presents an analytic-based scheme to obtain low WCETs based on the *Chebyshev theorem* at design-time. We also show the relationship between the low WCETs and mode switching probability, and formulate and solve the problem for improving resource utilization and reducing the mode switching probability. This approach sets the optimum static WCETs for HC tasks; however, tasks are rarely executed up to their WCETs at run-time. Therefore, to adapt dynamism at run-time, we propose a learning-based approach to consider the run-time behavior of tasks that dynamically monitors the tasks' execution times and adjusts the low WCETs to improve the QoS at the end of system execution. Further, we analyze the LC task dropping in the HI mode to improve QoS. We first propose a heuristic in which a new metric is defined that determines the number of allowable drops in the HI mode. Then, the task schedulability analysis is developed based on the new metric. Since the occurrence of the worst-case scenario at run-time is a rare event, a learning-based drop-aware task scheduling mechanism is then proposed, which carefully monitors the alterations in the behavior of the MC system at run-time to exploit the generated dynamic slacks for improving the QoS.

Another critical design challenge is how to improve QoS using the parallel feature of multi-core hardware platforms while managing the high power consumption and temperature of these platforms. We develop a tree of possible task mapping and scheduling at design-time (it would be exploited at run-time) to cover all possible scenarios of task overrunning and reduce the LC task drop rate in the HI mode while managing the power and temperature in each scenario of task scheduling. Since the dynamic slack is generated due to the early execution of tasks at run-time, we propose an online approach to reduce the power consumption and maximum temperature by using low-power techniques like dynamic voltage and frequency scaling and task re-mapping, while preserving the QoS. Specifically, our approach examines multiple tasks ahead (i.e., when a dynamic slack is generated) to determine the most appropriate task for the slack assignment that has the most significant effect on power consumption and temperature. However, changing the frequency and selecting a proper task for slack assignment and a suitable core for task re-mapping at run-time can be time-consuming and may cause deadline violation which is not admissible for HC tasks. Therefore, we analyze and then optimize the run-time scheduler and evaluate it for various platforms.

Acknowledgments

We would like to thank our group members and collaborators, especially Dr. Tuan Duy Anh Nguyen and Dr. Ali Hoseinghorban, for their support in realizing this work.

Contents

List of Acronyms

ACET	Average-Case Execution Time	37
AET	Actual Execution Time	25
CC	Cruise Controller	142
CDF	Cumulative Distribution Function	38
DAG	Directed Acyclic Graph	18
DVFS	Dynamic Voltage and Frequency Scaling	11
ER-EDF	Early Release Earliest Deadline First	23
EDF	Earliest Deadline First	112
EDF-VD	Earliest Deadline First with Virtual Deadline	23
EVT	Extreme Value Theory	25
FMS	Flight Management System	94
FP	Fixed Periority	23
GA	Genetic Algorithms	10
HC	High-Criticality	160
HI mode	HIgh-criticality mode	77
HP	Hyper Period	71
ILP	Integer Linear Programming	160
LC	Low-Criticality	77
LCM	Least Common Multiple	60
LO mode	LOw-criticality mode	3
MC	Mixed-Criticality	77
MILP	Mixed Integer Linear Programming	49
ML	Machine Learning	10
NDM	Number of Deadline Misses	115
PDF	Probability Density Function	44
PFH	Probability-of-Failure-per-Hour	18
PoF	Probability-of-Failure	20
QoS	Quality of Service	2
RL	Reinforcement Learning	63
RMSE	Root Mean Square Error	175
RTA	Response Time Analysis	96

SIL Safety Integrity Level 1
TDP Thermal Design Power 7
UAV Unmanned Air Vehicle 183
WCET Worst-Case Execution Time 2

Chapter 1
Introduction

Nowadays, embedded real-time systems have become significant in almost every aspect of industrial and human life, due to the increasing amount of computation parts in a small device. As a consequence of the ubiquity of embedded systems, they are often employed in safety-critical application domains, such as automotive, avionics, and medical devices. Autonomous driving is a part of the automotive domain, in which there are four primary functions—perception, planning, and decision, motion and vehicle control, and system supervision [1]. The perception stage is responsible for creating a reliable representation of the vehicle, where localization, mapping, and object detection functions are performed. These functions are real time, which means the correct output result of the functions must be ready in time (i.e., within specified time constraints defined for each function). For example, a vehicle control function that is responsible for steering, acceleration, and brake stroking must operate within its time constraint [2], or the obstacle detection function must perform complex sensing and estimations in real time to prevent serious damages like a fatal accident [1].

Several issues emerge in designing these safety-critical applications. Some critical issues that must be addressed are how to design such systems and guarantee timing, reliability, and safety requirements. Consider safety and reliability in autonomous vehicles, where various functions are incorporated along with safety requirements. An example of reliable behavior of an autonomous car is the desirable, reliable, and on-time behavior of the car to distinguish the obstacle on the road, pedestrians crossing the road, overtaking, and giving way. The safety and reliability of each function must meet the safety and reliability standards used in industries, like ISO26262 for road vehicles which is an extension of IEC 61508 [3]. These standards define different levels of safety for functions, called Safety Integrity Level (SIL) for automotive domains [4, 5], shown in Table 1.1. SIL is introduced in four levels in which SIL-1 has the lowest level of safety and SIL-4 has the highest level, where the ability to avoid harm or damage is more crucial in higher SIL. Probability-of-Failure-per-Hour (PFH) is a metric that is used for the

B. Ranjbar et al., *Quality-of-Service Aware Design and Management of Embedded Mixed-Criticality Systems*, https://doi.org/10.1007/978-3-031-38960-3_1

Table 1.1 IEC 61508 safety standard [9]

x	SIL-4	SIL-3	SIL-2	SIL-1
PFH_x	$<10^{-8}$	$<10^{-7}$	$<10^{-6}$	$<10^{-5}$
Failure condition	Catastrophic	Hazardous	Major	Minor

safety measurements of functions [6]. As shown, PFH has stricter constraints for higher criticality levels. The functions with different safety requirements must be executed and may communicate with other functions without sacrificing real-time and safety requirements. In conventional safety-critical real-time systems, tasks with the same criticality levels are executed on one hardware platform. Therefore, having multiple hardware platforms associated with multiple criticality levels would lead to high communication, space, and power consumption. As a result, MC systems have emerged as an effective solution in various industries, where multiple tasks with different criticality levels are executed on a common hardware platform in order to meet requirements such as cost, space, weight, power consumption, and communication while guaranteeing a safe operation. The criticality of the tasks is based on their importance and functionality for the application [7, 8].

The main research question in designing these MC systems is how to reconcile the conflicting requirements of ensuring safety and real-time constraints and sharing for efficient resource usage on a common platform. A lot of progress has been made in both academic and industrial aspects since 2007, especially in the last decade, to design, model, manage, implement, and evaluate these MC systems [10]. In these MC systems, tasks have to be analyzed at design-time to obtain their parameters, like WCET [7, 10]. As Burns and Davis mentioned in [10], how the WCETs are computed is one of the challenges in modeling and designing the MC systems. Then, by employing these task parameters, proper MC task scheduling strategies are derived to satisfy the safety and real-time constraints and optimize the processor capacity usage [11]. In order to guarantee the real-time constraints of tasks with HC levels, some/all tasks with Low-Criticality (LC) levels might be dropped/degraded in some situations in favor of HC tasks [10]. Therefore, the QoS (the percentage of executed LC tasks to all LC tasks) should be improved at design- and run-time from the MC task modeling and scheduling perspectives.

In addition, MC systems are getting more complicated due to ever-increasing computational requirements and the growth in the number of tasks; therefore, multi-core platforms are utilized to execute the tasks in parallel, thereby improving the system performance [11]. As the degree of freedom (in terms of availability of the cores) increases, power consumption and high temperature are issues of crucial importance in MC systems. It is not trivial to guarantee the real-time constraints and improve the QoS while managing the system power consumption. Systems with high peak power consumption are more likely to generate unexpected heat beyond the intended cooling capacity. These systems will be more susceptible to failures and instability [12]. In other words, the reliability, lifetime, and timeliness of these systems will be undesirably affected [13]. As a result, minimizing power consumption in multi-core MC systems while ensuring the real-time constraints and guaranteeing the minimum QoS is a significant issue that should be addressed.

This book addresses the challenges associated with the design and model of MC applications and the management of MC systems on multi-core platforms. We present an offline theoretical-based scheme and an online adaptive scheme in this book to determine the MC application's parameters and improve the QoS. Moreover, we propose a parameter for each task in order to improve the QoS and, based on the introduced parameter, develop a design-time schedulability analysis and a run-time task scheduler aiming at QoS improvement. Finally, by considering the MC hardware aspect, it presents novel approaches to managing peak power consumption and maximum temperature in multi-core MC systems at both design- and run-time while ensuring the real-timeliness and improving the QoS.

The rest of this chapter is organized as follows. In Sect. 1.1, we provide a summary of MC system definition and properties. We look in-depth at the MC system design and modeling from the applications perspective and their issues. Section 1.2 presents the trends in MC hardware design and management when exploiting the multi-core platforms. Then, in Sect. 1.3, we introduce the research questions and summarize the research challenges that need to be solved. Section 1.4 presents the book contributions and in the end, Sects. 1.5 and 1.6 outline the book, and their organization, respectively.

The content of this book is based on [14].

1.1 Mixed-Criticality Application Design

In most of the safety-critical real-time applications (in medical, flight control, etc. devices), tasks are classified into multiple criticality levels in order to maintain the predictability of the applications under different unexpected behaviors. In these real-time systems, these tasks have to be analyzed at design-time to obtain their WCET and then are scheduled based on their obtained WCET [7]. Many approaches like those presented in [15, 16] and tools like OTAWA [17] are used to determine the high WCET of a task by analyzing the task's control flow graph. These tools provide a safe and conservative execution time-bound so that no task's execution time exceeds the WCET under any circumstances. However, Fig. 1.1 [15] shows an execution time distribution of a task and observes that most samples' execution time is significantly shorter than such a conservative WCET. As a result, the resources would be severely underutilized at run-time, which leads to poor processor utilization and QoS in conventional real-time systems.

To this end, MC systems are designed to tackle this issue, where tasks are analyzed with different assumptions—for example, optimistic and pessimistic assumptions for a system in which two criticality levels of tasks are executed—to obtain multiple WCETs, corresponding to the multiple criticality levels and the operation mode of the system [18–20]. This ensures that the processor utilization (and correspondingly, the QoS) is maximized in the LOw-criticality mode (LO mode),

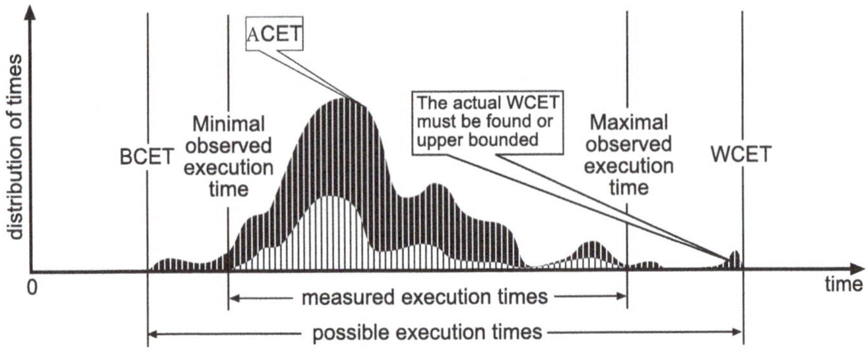

Fig. 1.1 Execution time distribution for a real-time task [15]

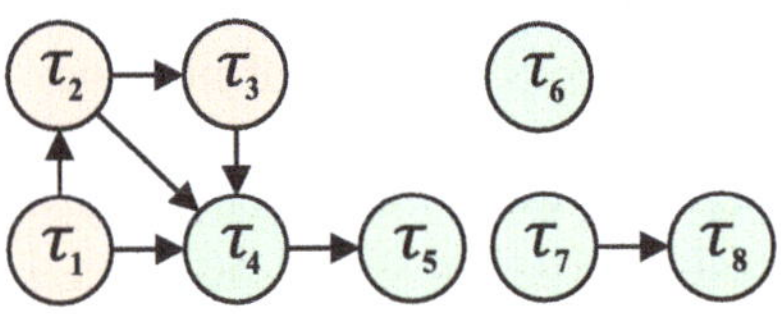

	τ_1	τ_2	τ_3	τ_4	τ_5	τ_6	τ_7	τ_8
$WCET_i^{LO}$	30	60	20	20	30	60	20	20
$WCET_i^{HI}$	30	80	50	20	30	60	20	20

Fig. 1.2 An example of real-life application (unmanned air vehicle) task graph

while the guarantees are preserved in the HIgh-criticality mode (HI mode). Figure 1.2 shows a task set of Unmanned Air Vehicle (UAV), which is a real-life MC application [21]. This application is composed of eight tasks, in which HC tasks have two different values of WCETs. Tasks τ_1 to τ_3 are HC tasks that are responsible for the avoidance, navigation, and stability of the system, respectively. Failure in the execution of these tasks before their deadline may lead to a system failure and irreparable damage to the system. The roles of LC tasks (τ_4 to τ_8) are recording sensor data, GPS coordination, and video transmissions, which help the system to carry out its mission successfully [21]. From the MC task scheduling perspective, tasks are first scheduled in the LO mode based on their low WCETs. At run-time, if the execution time of at least one HC task exceeds its low WCET (a task overruns), the system switches to the HI mode, i.e., the *mode switch* occurs due to the HC tasks' overrunning. In HI mode, to guarantee the correct execution of HC tasks, the system switches to the second schedule, where all HC tasks are scheduled based on their high WCETs and all or some LC tasks are dropped, which leads to QoS degradation [7, 10, 18, 22–24].

When the gap between the low and high WCETs is large, more tasks, especially LC tasks, are guaranteed to be scheduled in a processor at design-time. However, it may cause frequent system mode switches and, consequently, drop more LC tasks at run-time due to inefficient low WCET determination for HC tasks. When this gap is small, fewer LC tasks are scheduled in the LO mode which underutilizes the processor. Indeed, this is overly pessimistic because, as shown in Fig. 1.1, tasks would be executed with less likelihood up to observed or actual WCET.

Therefore, low WCETs play an important role in designing efficient MC systems and improving the timing behavior of these systems.

Nevertheless, according to the limitation of an MC system formal model, which is mentioned in [4, 5], dropping LC tasks (except for non-criticality tasks) in favor of HC tasks is not a suitable protection mechanism in industrial applications. The frequent deadline misses or service degradation of some LC tasks, such as mission-critical tasks, may have a negative impact on the other HC tasks and mission-critical tasks themselves, and consequently on the entire system, and may prevent the system from accomplishing its mission correctly. For instance, consider an MC system whose mission is to capture images in specific time intervals. In this application, the engine operation (i.e., the function that ensures the safe execution of the operation) and the operation of capturing images are considered as the HC tasks and LC tasks, respectively [25, 26]. Accordingly, if the system switches to the HI mode, due to the execution of HC tasks, the existing scheduling methods may frequently drop LC tanks or suspend them for a long time, which is not acceptable for a system whose main mission is capturing images. As can be realized, dropping LC tasks (except for non-criticality tasks) is not permitted in industrial applications, but depending on the type of the application, some of the LC tasks (i.e., mission-critical tasks) could be dropped. Therefore, since the frequent dropping or postponing of their execution for a long time in the HI mode is not appropriate, how to model MC tasks, and design an MC system, is crucial in improving the QoS.

As most MC applications are safety-critical, the system must be designed carefully at design-time to execute all HC tasks correctly before their deadlines, even in worst-case situations, e.g., tasks are executed up to their WCETs or the system would be in overload situation most of the time. For example, if the system switches to the HI mode most of the time due to HC task overrunning, the minimum QoS must be guaranteed at design-time to ensure that the system can perform its mission effectively. Most MC systems are designed statically at design-time to guarantee the worst-case scenario. However, the system does not always exhibit the worst-case behavior at run-time. For example, much dynamic slack (the time between a task's actual completion time and its WCET) would be generated at run-time due to the early finish of task execution, which can be used to improve the objectives. Consequently, these systems cannot adapt to task dynamism at run-time, which results in significant performance losses for LC tasks. In this book, dynamism occurs when the tasks' execution times are varied due to changes in tasks' inputs. Figure 1.3 depicts measured radiation dose in an airplane with varying altitudes. As shown, the radiation during a flight increases with the airplane's altitude (i.e., in the operating environment), and the maximum radiation is almost 20X as compared to when the aircraft is on the ground. The system is designed to operate in the worst-case scenario, i.e., when it is in the air, and the radiation has the maximum value. Although the airplane is in the air for the bulk of the flight time, which has the maximum radiation value, the system should support other tasks during landing (when the radiation has a low value), which is more critical. Besides, the variations in the operating environment could generate different inputs for tasks and, therefore, may lead to the computational demand being higher than the processor capacity

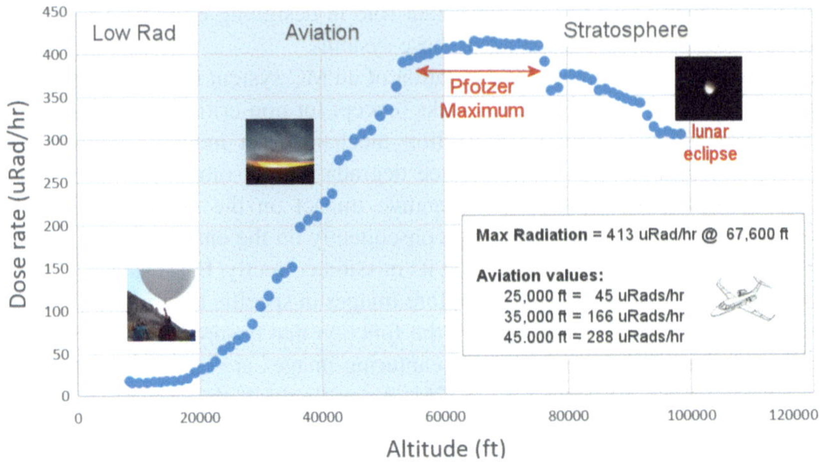

Fig. 1.3 The relation between radiation and altitude of an airplane due to varying operating environment [27]

under the specified timing requirements. Thus, the MC system's run-time behavior can be investigated in addition to system design in design-time phase to improve the objectives like QoS while ensuring the real-timeliness based on the dynamic changes during run-time.

1.2 Mixed-Criticality Hardware Design

A large number of real-time systems are embedded in small battery-operated devices. In these systems, efficient power management is vital for achieving the performance desired in these systems [28, 29]. Besides, in general, in some embedded real-time systems, applications consist of various control tasks that execute together to achieve a common goal like controlling autonomous driving. In other words, tasks are dependent, and there are precedence relations between tasks in such systems [28]. Therefore, since there are a large number of tasks in these sophisticated embedded systems to be executed and communicate with each other on a single platform to meet cost, power, and performance, the platforms are migrating from single cores to multi-cores/many cores. The multi-core platforms can be utilized to deal with high-performance requirements and to improve the QoS by efficiently allocating tasks among cores.

In these multi-core MC systems, platforms require higher power to operate, compared to single-core platforms, mainly when the system is in the HI mode. The reason is that, by executing all/most HC tasks in this situation, there may be an increase in computational demands, and the system may then become overloaded [30]. Thus, all cores may execute tasks simultaneously to meet the deadlines of tasks, which increases the instantaneous processor power beyond its Thermal Design Power (TDP) constraint [12, 13, 31]. TDP is the maximum sustainable power that a chip can dissipate safely. If the task scheduler is not aware of the power consumption, all cores might be activated simultaneously with the highest performance. Therefore, the system will draw a significantly larger power than it is designed for. Systems with high peak power are more likely to generate unexpected heat beyond the chip's intended cooling capacity. They will be more susceptible to failures and instability [12], which is not acceptable for the HC tasks, and it may cause catastrophic consequences. In addition, as the degree of freedom (in terms of the availability of the cores) increases, it is not trivial to guarantee real-time constraints while managing the system's peak power. In other words, the reliability, lifetime, and timeliness of these systems will be adversely affected [13]. Therefore, managing the peak power consumption and maximum temperature of the multi-core system, while guaranteeing the deadlines of tasks at runtime, is crucial to be studied.

In multi-core platforms, all cores being active simultaneously and executing all tasks with their pessimistic power consumption value up to their WCET values might be the worst-case scenario from a power consumption perspective. As mentioned in the previous section, this worst-case scenario must meet the deadline constraints and guarantee the minimum QoS. However, all cores in multi-core MC systems may not always be in the worst case of their mission. A run-time management policy is needed to adapt the system to dynamic changes, like employing the dynamic slack on all cores to reduce the power consumption and maximum temperature and improve the QoS.

To summarize, the following are the trends and requirements we observe in MC hardware design:

- The need and capability for concurrent task execution in a multi-core platform are increasing.
- Concerns over power consumption in multi-core platforms are growing more serious in MC systems.
- The high processor temperature is becoming an increasingly important concern for the correct and safe execution of MC applications on platforms.

1.3 Research Challenges and Questions

The designers face challenges while designing MC systems at low space, cost, and power. In order to keep the space low, the system resources need to be used efficiently. The important design points and trends mentioned in Sects. 1.1 and 1.2, respectively, indicate the increasing complexity of MC system design and management at both design- and run-time. Burns and Davis [10] provide the challenges and open issues in the field of MC systems. However, the following are the major challenges in the design, analysis, and management of embedded MC systems, which we discussed in the previous sections and deal with in this book.

- Analyzing and adjusting the WCETs for HC tasks to be employed in MC task scheduling
- Investigating the MC system behavior in the HI mode and determining the number of allowable drops for LC tasks
- Integrating and adapting the MC systems with the run-time behavior of applications
- Improving the QoS in multi-core platforms by executing the tasks in parallel
- Managing the power and maximum temperature in multi-core MC systems while guaranteeing the real-time constraints of applications

According to the discussions in the previous sections and the challenges faced by this book, the following research objective are addressed in this book:

Modeling, designing, and management of embedded MC system to improve the QoS while ensuring the real-time and safety constraints of functions

In order to achieve the above objective, the following research questions (RQ) must be answered while designing the MC systems, analyzing MC applications, and deploying the MC applications on multi-core platforms:

RQ1: How can we obtain/estimate the safe and tight low WCET for HC tasks in order to improve QoS (i.e., reducing the number of dropped LC tasks to the maximum extent possible) and manage the mode switches' probability?

RQ2: How can the MC systems adapt to dynamism (i.e., run-time task behavior) at run-time in order to improve QoS and manage the mode switches' probability?

RQ3: How can we model the MC tasks, and regarding this model, how is the task schedulability tested in order to reduce the possibility of frequent drops of LC tasks in the HI mode and improve the QoS?

RQ4: How can the run-time behavior help the task scheduler, with the low overhead in a way that drops fewer LC tasks in the HI mode, while guaranteeing the real-time constraints?

RQ5: How can the hardware parallelism feature of multi-core platforms be employed for MC applications in order to improve the QoS while guaranteeing the real-time constraints?

RQ6: How can we overcome the power consumption and thermal issues in multi-core platforms with low timing and memory overheads while managing the QoS, real-timeliness, and safety?

RQ7: Which system-level low-power techniques can be employed in embedded MC systems in order to manage both maximum temperature and power consumption?

RQ8: How can the generated dynamic slack be efficiently employed for various system objective improvements like decreasing power consumption, postponing the possibility of mode switches, and improving the QoS of LC tasks?

1.4 Key Contributions

In order to answer the research questions, this book has mainly focused on how to model and schedule the MC applications and how to exploit the multi-core platform features to design and manage the MC systems. Figure 1.4 shows the overall structure of the book to address the mentioned research challenges and questions. We first focus on application-level analysis which can be applied to any hardware, single or multi-core platforms. Nevertheless, we consider independent MC tasks, executing on a single-core platform, to achieve the objective of analyzing, modeling, and designing the MC systems. We propose two novel contributions of QoS improvement through WCET analysis and task dropping analysis and modeling. We investigate and solve the research problem at design- and run-time for both contributions.

As mentioned in Sect. 1.2, in some embedded real-time systems, MC applications consist of dependent tasks, in which a large number of functions execute on a common multi-core platform. To this end, we consider the dependent task model that the tasks are executed on multi-core platforms. We employ the hardware parallelism feature of multi-core platforms to design MC systems and improve the QoS. In such multi-core platforms, we also address the power and thermal issues, which may lead to an unsafe point in MC system design. In the following, we detail the contributions mentioned in this figure.

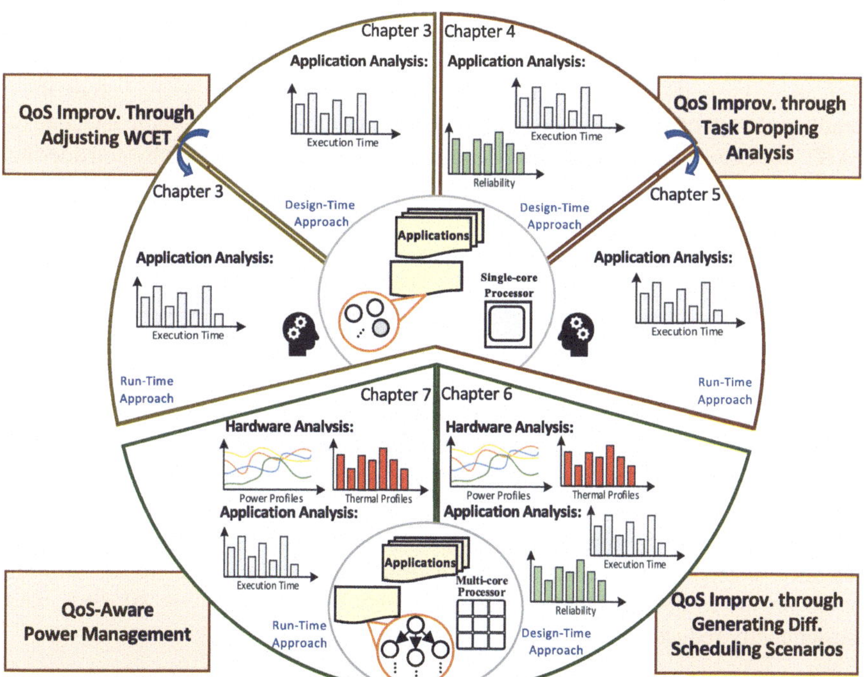

Fig. 1.4 Overall structure of the book

1.4.1 Application Analysis and Modeling

In order to address the challenge of WCET adjustment of HC tasks to be employed in the LO mode, we first address the RQ1 by analyzing the low WCET parameter and determining the appropriate values for MC tasks. We present the novel analytical scheme, called *BOT-MICS*, to provide a reasonable trade-off between the number of LC tasks that can be guaranteed to meet the deadlines at design-time (i.e., QoS) and the probability of mode switching at run-time. In this design-time approach, the WCETs are obtained based on the *Chebyshev theorem*, and then we show the relation between the low WCETs and mode switching probability. The Genetic Algorithms (GA) is used to formulate and solve the optimization problem for improving the resource utilization and reducing the mode switching probability.

Then, to address the intended objective and answer RQ2, an online scheme, called *ADAPTIVE*, is proposed to be employed at run-time. It studies and analyzes the run-time behavior of task execution and presents a dynamic QoS-aware scheduling algorithm. This algorithm adjusts the low WCET of HC tasks based on the available accumulated dynamic slack at run-time to improve the results' quality based on the system changes while guaranteeing the minimum service of LC tasks, even in the HI mode by considering a utilization threshold when adjusting the low WCETs.

We further address the challenge of drop-aware behavior analysis of MC systems to answer RQ3. We introduce *FANTOM*, a heuristic, in which a new task parameter is defined, and then based on the defined parameter, schedulability analysis of MC tasks is developed by considering safety requirements. In FANTOM, the schedulability analysis is conducted in an offline manner in order to guarantee that all tasks with different criticality levels are executed properly before their deadlines based on the operational mode of MC systems. Thus, the main objective of FANTOM is to execute the majority of the LC tasks in the HI mode by considering a maximum allowable number of drops for every LC task.

The offline techniques, like *FANTOM*, are mostly pessimistic, as the occurrence of the worst-case scenario at run-time is rare. Therefore, we propose a novel optimistic mechanism that reduces the number of drops for the LC tasks by observing the system's behavioral changes at run-time. The answer of RQ4 has been achieved by exploiting the generated dynamic slacks in the decision-making process for the online task dropping to execute more LC tasks in the HI mode and enhance their schedulability. Since we are unaware of the amount of generated dynamic slacks during run-time in advance, Machine Learning (ML) approaches can be employed as a management technique for the prediction. Therefore, utilizing ML techniques as part of the proposed approach has enabled it to partially exploit the dynamic slack to improve the QoS for the LC tasks in the HI mode. In these schemes, the learner finds the optimum drop rate for the LC tasks at run-time based on the available dynamic slack, prevents frequent drops in HI mode, and consequently reduces their deadline miss rate.

1.4.2 Multi-core Mixed-Criticality System Design

Now, we address the challenge of using the parallel execution of tasks in multi-core platforms to achieve the research objective and answer RQ5 and RQ6. We first propose a method that exploits a tree of schedules for dependent MC tasks running on multi-core platforms. This tree of schedules is generated at design-time, considering system safety (i.e., all possibilities of fault occurrence scenarios in different tasks) and task overrun. When an HC task overruns or a fault occurs at run-time, the scheduler chooses the proper schedule from the tree to tolerate the faults or manage the system mode switches with low overheads. Our technique aims to improve the LC tasks' QoS in the HI mode while all HC tasks meet their deadlines. Besides, high-power consumption and temperature are crucial issues in MC systems while using multi-core platforms. As a result, by generating the schedule tree and exploiting it at run-time, the LC tasks' QoS is maximized in each schedule of the tree while managing the system's peak power consumption and tolerating the occurrence of possible faults.

We further exploit the run-time execution feature to address the research challenges in executing MC applications on multi-core platforms. Therefore, the RQ7 is first needed to be answered. Then, based on the study, we propose a heuristic to manage power consumption in MC systems during run-time. To achieve this, we exploit dynamic slacks, the slack between tasks' actual completion time and their WCET, along with Dynamic Voltage and Frequency Scaling (DVFS), a system-level low-power technique. Our approach has two phases: (1) at design-time, the tasks are scheduled on each core, and the resulting schedule is stored to be used as a static scheduling table. This is performed for both LO mode and HI mode, defined in Sect. 1.1. In this case, the number of LC tasks that have to be dropped in the HI mode is minimized to improve the system's overall QoS. (2) At run-time, in order to answer RQ8, we examine multiple tasks in the future (look-ahead) to select the most appropriate task to assign the currently available dynamic slack. The selection is based on the impact of the tasks on the power consumption and temperature of the system, which is quantified by a weighted multi-objective cost function. Therefore, the speed of the core that runs the task can be decreased accordingly. Additionally, besides exploiting the dynamic slacks, we propose a task re-mapping technique (as a low-power technique) at run-time to improve the system temperature profile further. However, the online scheduler's timing overhead to select an appropriate task and check the re-mapping technique to choose a proper core are crucial for the MC systems and may cause deadline violations. Furthermore, the timing overhead of changing *V-f* levels in using the DVFS technique is critical in run-time task scheduling. Therefore, to answer RQ6 in this chapter from the timing overhead perspective, we analyze and evaluate the effect of these overheads on the schedule of MC tasks in real multi-core platforms.

1.5 Book Outline

The remainder of this book is organized as follows:

Chapter 2 "Preliminaries and Related Work" presents the concept and required preliminaries to understand the modeling and task scheduling. Then, a survey of related works in the domain under consideration, like MC task scheduling mechanisms, QoS improvement, and QoS-aware power management in MC systems, is reviewed in this chapter. Then, to highlight the contributions made, this book is divided into five chapters.

Chapter 3 "Bounding Time in MC Systems" presents the novel schemes to adjust the low WCET of MC tasks in order to improve QoS and reduce the mode switching probability. This chapter proposes a design-time analytical approach and an adaptive run-time approach. This chapter is based on [32, 33] and [34, 35].

Chapter 4 "Safety- and Task-Drop-Aware MC Task Scheduling" proposes a heuristic by first defining a new parameter for each task to improve the QoS by introducing a maximum allowable number of drops for every LC task. Then, based on the defined parameter, an MC task schedulability analysis is developed by considering safety requirements. This chapter is based on [36].

Chapter 5 "Learning-Based Drop-Aware MC Task Scheduling" provides an adaptive run-time scheme, where a learning-based drop-aware MC task scheduling mechanism is proposed to improve the QoS by exploiting the generated dynamic slacks. This chapter is based on [37].

Chapter 6 "Fault-Tolerance- and Power-Aware Multi-core MC System Design" proposes a design-time QoS-aware power management in multi-core MC systems. In this chapter, a design-time approach is presented in order to improve QoS by generating different scheduling scenarios while reducing power consumption. This chapter is based on [38].

Chapter 7 "QoS- and Power-Aware Run-Time Scheduler for Multi-core MC Systems" presents a run-time QoS-aware approach to manage power consumption and maximum temperature by exploiting the generated dynamic slacks. This chapter is based on [39] and [40].

Chapter 8 "Conclusions and Future Work" concludes the book and presents a brief discussion of possible future research works within the domain.

1.6 Conclusions

Nowadays, implementing a complex system, and executing various applications with different levels of assurance, is a growing trend in modern embedded real-time systems, which are known as MC systems. In these systems, a deadline miss in tasks with various criticality levels has a different consequences on the system, from no impact to catastrophic consequences. Therefore, an efficient MC system design should be developed to guarantee the safe execution of HC tasks to prevent catastrophic damages while improving the QoS. In order to design, analyze, and manage an efficient MC system, there are several major challenges, which are discussed in this chapter.

In order to address the challenges, we focused on application- and hardware-level analysis. In application-level analysis, two novel approaches for QoS improvement through WCET analysis and task dropping analysis are proposed for both design-time and run-time phases. In addition, in hardware analysis, the hardware parallelism of multi-core platforms is employed in designing an MC system at design- and run-time, in order to improve the QoS, while addressing the power and thermal issues.

References

1. De Jong Yeong et al. "Sensor and Sensor Fusion Technology in Autonomous Vehicles: A Review". In: *Sensors* 21.6 (2021). DOI: `10.3390/s21062140`.
2. Shinpei Kato et al. "An Open Approach to Autonomous Vehicles". In: *IEEE Micro* 35.6 (2015), pp. 60–68. DOI: `10.1109/MM.2015.133`.
3. ISO 26262. *Road vehicles—Functional safety*. Standard. International Organization for Standardization (ISO), Dec. 2018.
4. Alexandre Esper et al. "How realistic is the mixed-criticality real-time system model?" In: *Proc. of Real-Time Networks and Systems (RTNS)*. ACM. 2015, pp. 139–148.
5. R. Ernst and M. Di Natale. "Mixed Criticality Systems—A History of Misconceptions?" In: *IEEE Design & Test* 33.5 (2016), pp. 65–74.
6. P. Huang, H. Yang, and L. Thiele. "On the scheduling of fault-tolerant mixed-criticality systems". In: *Proc. on Design Automation Conference (DAC)*. 2014, pp. 1–6.
7. S. Baruah et al. "The preemptive uniprocessor scheduling of mixed-criticality implicit-deadline sporadic task systems". In: *Proc. of Euromicro Conference on Real-Time Systems (ECRTS)*. 2012, pp. 145–154.
8. Hang Su and Dakai Zhu. "An elastic mixed-criticality task model and its scheduling algorithm". In: *Proc. on Design, Automation & Test in Europe Conference & Exhibition (DATE)*. IEEE. 2013, pp. 147–152.
9. William M Goble. *Control systems safety evaluation and reliability*. ISA, 2010.
10. Alan Burns and Robert I. Davis. "A Survey of Research into Mixed Criticality Systems". In: *ACM Computing Surveys (CSUR)* 50.6 (2017), pp. 1–37.
11. M. A. Awan, D. Masson, and E. Tovar. "Energy efficient mapping of mixed criticality applications on unrelated heterogeneous multicore platforms". In: *Proc. on IEEE Symposium on Industrial Embedded Systems (SIES)*. 2016, pp. 1–10.
12. Waqaas Munawar et al. "Peak Power Management for scheduling real-time tasks on heterogeneous many-core systems". In: *Proc. of the International Conference on Parallel and Distributed Systems (ICPADS)*. 2014, pp. 200–209.
13. Jinkyu Lee, Buyoung Yun, and Kang G Shin. "Reducing peak power consumption inmulticore systems without violating real-time constraints". In: *IEEE Transactions on Parallel and Distributed Systems (TPDS)* 25.4 (2014), pp. 1024–1033.
14. Behnaz Ranjbar. "Quality-of-Service Aware Design and Management of Embedded Mixed-Criticality Systems". PhD thesis. Technische Universität Dresden, 2022.
15. Reinhard Wilhelm et al. "The Worst-Case Execution-Time Problem—Overview of Methods and Survey of Tools". In: *ACM Trans. Embed. Comput. Syst. (TECS)* 7.3 (May 2008).

16. A. Kumar et al. "Iterative Probabilistic Performance Prediction for Multi-Application Multiprocessor Systems". In: *IEEE Transactions on Computer-Aided Design of Integrated Circuits and Systems (TCAD)* 29.4 (2010), pp. 538–551.
17. Clément Ballabriga et al. "OTAWA: an open toolbox for adaptive WCET analysis". In: *IFIP International Workshop on Software Technolgies for Embedded and Ubiquitous Systems*. Springer. 2010, pp. 35–46.
18. D. Liu et al. "EDF-VD Scheduling of Mixed-Criticality Systems with Degraded Quality Guarantees". In: *Proc. on IEEE Real-Time Systems Symposium (RTSS)*. 2016, pp. 35–46.
19. D. Liu et al. "Scheduling Analysis of Imprecise Mixed-Criticality Real-Time Tasks". In: *IEEE Transactions on Computers (TC)* 67.7 (2018), pp. 975–991.
20. G. Chen et al. "Utilization-Based Scheduling of Flexible Mixed-Criticality Real-Time Tasks". In: *IEEE Transactions on Computers (TC)* 67.4 (2018), pp. 543–558.
21. R. Medina, E. Borde, and L. Pautet. "Availability enhancement and analysis for mixed-criticality systems on multi-core". In: *Proc. on Design, Automation & Test in Europe Conference & Exhibition (DATE)*. 2018, pp. 1271–1276.
22. Z. Guo et al. "Uniprocessor Mixed-Criticality Scheduling with Graceful Degradation by Completion Rate". In: *Proc. on IEEE Real-Time Systems Symposium (RTSS)*. 2018, pp. 373–383.
23. Chuancai Gu et al. "Partitioned mixed-criticality scheduling on multi-processor platforms". In: *Proc. on Design, Automation & Test in Europe Conference & Exhibition (DATE)*. 2014, p. 292.
24. Siva Satyendra Sahoo, Behnaz Ranjbar, and Akash Kumar. "Reliability-Aware Resource Management in Multi-/Many-Core Systems: A Perspective Paper". In: *Journal of Low Power Electronics and Applications* 11.1 (2021), p. 7.
25. Sanjoy Baruah, Haohan Li, and Leen Stougie. "Towards the design of certifiable mixed-criticality systems". In: *Proc. of Real-Time and Embedded Technology and Applications Symposium (RTAS)*. IEEE. 2010, pp. 13–22.
26. Sanjoy K Baruah et al. "Mixed-criticality scheduling of sporadic task systems". In: *European Symposium on Algorithms*. 2011, pp. 555–566.
27. TONY PHILLIPS. *Flying at Night Doesn't Protect You from Cosmic Rays*. http://Spaceweather.com. Accessed: June 2022. 2015.
28. Giorgio C Buttazzo. *Hard real-time computing systems: predictable scheduling algorithms and applications*. Vol. 24. Springer Science & Business Media, 2011.
29. M. Salehi and A. Ejlali. "A Hardware Platform for Evaluating Low-Energy Multiprocessor Embedded Systems Based on COTS Devices". In: *IEEE Transactions on Industrial Electronics (TIE)* 62.2 (2015), pp. 1262–1269.
30. James H Anderson, Sanjoy K Baruah, and Björn B Brandenburg. "Multicore operating-system support for mixed criticality". In: *Proc. of the Workshop on Mixed Criticality: Roadmap to Evolving UAV Certification*. Vol. 4. Citeseer. 2009, p. 7.
31. BongKi Lee et al. "Peak power reduction methodology for multi-core systems". In: *Proc. of the International SoC Design Conference (ISOCC)*. 2010, pp. 233–235.
32. B. Ranjbar et al. "Improving the Timing Behaviour of Mixed-Criticality Systems Using Chebyshev's Theorem". In: *Proc. on Design, Automation & Test in Europe Conference & Exhibition (DATE)*. 2021, pp. 264–269.
33. Behnaz Ranjbar et al. "BOT-MICS: Bounding Time Using Analytics in Mixed-Criticality Systems". In: *IEEE Transactions on Computer-Aided Design of Integrated Circuits and Systems (TCAD)* 41.10 (2022), pp. 3239–3251. DOI: 10.1109/TCAD.2021.3127867.
34. B. Ranjbar, A. Hoseinghorban, and A. Kumar. "Motivating Agent-Based Learning For Bounding Time in Mixed-Criticality Systems". In: *Proc. on Design, Automation & Test in Europe Conference & Exhibition (DATE)*. 2023.
35. B. Ranjbar, A. Hoseinghorban, and A. Kumar. "ADAPTIVE: Agent-Based Learning for Bounding Time in Mixed-Criticality Systems". In: *Proc. on Design Automation Conference (DAC)*. 2023.

36. B. Ranjbar et al. "FANTOM: Fault Tolerant Task-Drop Aware Scheduling for Mixed-Criticality Systems". In: *IEEE Access* 8 (2020), pp. 187232–187248. DOI: 10.1109/ACCESS.2020.3031039.
37. Behnaz Ranjbar et al. "Learning-Oriented QoS- and Drop-Aware Task Scheduling for Mixed-Criticality Systems". In: *Computers* 11.7 (2022). DOI: 10.3390/computers11070101.
38. Behnaz Ranjbar et al. "Toward the Design of Fault-Tolerance-Aware and Peak-Power-Aware Multicore Mixed-Criticality Systems". In: *IEEE Transactions on Computer-Aided Design of Integrated Circuits and Systems (TCAD)* 41.5 (2022), pp. 1509–1522. DOI: 10.1109/TCAD.2021.3082495.
39. B. Ranjbar et al. "Online Peak Power and Maximum Temperature Management in Multi-core Mixed-Criticality Embedded Systems". In: *Proc. of Euromicro Conference on Digital System Design (DSD)*. 2019, pp. 546–553.
40. Behnaz Ranjbar et al. "Power-Aware Runtime Scheduler for Mixed-Criticality Systems on Multicore Platform". In: *IEEE Transactions on Computer-Aided Design of Integrated Circuits and Systems (TCAD)* 40.10 (2021), pp. 2009–2023. DOI: 10.1109/TCAD.2020.3033374.

Chapter 2
Preliminaries and Related Work

The previous chapter presented an introduction to the trends and issues in QoS-aware MC system design. In designing such systems, scheduling algorithms are used and need to satisfy all timing constraints corresponding to each operational mode when used in MC systems. In order to accomplish this work, first, the application, hardware, and power need to be appropriately modeled. Then, the scheduling algorithms need to be precisely selected and employed in these MC systems. The chapter mainly introduces the relevant preliminaries and common models of MC systems, which are employed in this book. A brief overview of the related literature works is also provided.

This chapter is organized as follows. Section 2.1 presets the preliminaries used in this book, in which first a brief introduction of MC systems, like MC application model, QoS definition, and system operational model, is presented in Sect. 2.1.1. Then, we provide a brief overview of the fault model, used fault-tolerance techniques, and safety requirements in Sect. 2.1.2. The hardware architecture model and power consumption model are presented in Sects. 2.1.3 and 2.1.4, respectively. The second section of this chapter (Sect. 2.2) provides an overview of the related state-of-the-art research works, where first a survey of MC task scheduling algorithms is presented in Sect. 2.2.1 along with the chosen scheduling algorithm in this book. The relevant related works in improving the field of MC system timing behavior through WCET adjustment and task dropping analysis are provided in Sect. 2.2.2. At the end of the section, we overview the related publications in the field of MC system hardware design and the considered challenges in Sect. 2.2.3. Finally, we conclude the chapter in Sect. 2.3.

B. Ranjbar et al., *Quality-of-Service Aware Design and Management of Embedded Mixed-Criticality Systems*, https://doi.org/10.1007/978-3-031-38960-3_2

2.1 Preliminaries

2.1.1 Mixed-Criticality Systems

In the following, we present the MC task model and characteristics of a task, QoS definition, and system operational model.

2.1.1.1 Mixed-Criticality Application Model and Specification

Analogous to most state-of-the-art works, we consider real-time applications consisting of n periodic MC tasks $\{\tau_1, \tau_2, \ldots, \tau_n\}$, such that each task τ_i is characterized as:

$$\tau_i = (\zeta_i, WCET_i^{LO}, WCET_i^{HI}, d_i, T_i) \tag{2.1}$$

where

- ζ_i denotes the criticality level of τ_i.
- $WCET_i^{LO}$ ($WCET_i^{HI}$) denotes the WCET of task τ_i in LO mode (HI mode).
- T_i denotes the period of task τ_i, which is the minimum amount of the time between two released instances.
- d_i denotes the deadline of task τ_i.

Some MC systems require a high level of safety due to their timing requirements [1]. We have exploited criticality levels similar to what was defined in [2], in which each criticality level has a requirement based on the deadline and safety requirements. Table 2.1 represents an industrial standard safety requirements, e.g., DO-178B [3], which introduces five levels of safety, i.e., *A*, *B*, *C*, *D*, and *E* (A and E provide the highest and the lowest levels of safety, respectively) [2, 4, 5]. As shown in this table, the occurrence of a failure in tasks with various criticality levels has a different impact on the system [6, 7]. To guarantee the system's safety, the Probability-of-Failure-per-Hour (PFH) (which is adopted by safety standards) is determined for all the criticality levels [5, 8], which is discussed in detail later in Sect. 2.1.2. Analogous to [9, 10], we consider dual-criticality system where each MC task can be either high-critical (ζ_i= HC) or low-critical (ζ_i= LC). Besides, due to having two different WCET values, for each Low-Criticality (LC) task $WCET_i^{LO} = WCET_i^{HI}$ and also, for each High-Criticality (HC) task $WCET_i^{LO} \leq WCET_i^{HI}$. In this book, we assume preemptive execution for tasks, which means the tasks are interrupted during their execution on a core that are mapped on it.

Depending on whether the MC tasks are dependent or independent, the task deadlines and periods are valued differently. In the case of independent tasks, analogous to state-of-the-art works like [11], the deadline of a task (d_i) is equal to its period (T_i). If the tasks are dependent (the task model is Directed Acyclic Graph

Table 2.1 DO-178B safety requirement [3]

x	A	B	C	D	E
PFH_x	$< 10^{-9}$	$< 10^{-7}$	$< 10^{-5}$	$> 10^{-5}$	-
Failure condition	Catastrophic	Hazardous	Major	Minor	No effect

(DAG), like what has been considered in the task model of [9, 10, 12]), each task has an identical period (T_i), which is equal to the period of the application (T_{app}) [13]. In dependent task models, each task has some successors and predecessors, which are determined by Su_i and Pr_i, respectively. A task is released and ready to be executed if all its predecessor tasks have finished their execution. The deadline and period of the task graph are equal for an application ($d_{app} = T_{app}$), but, for each task, a deadline d_i (which can be named as a local deadline in a task graph model) is determined in order that all its successors can be scheduled before their deadlines. Hence, the deadlines of tasks that have no successors are equal to the task graph's deadline. Besides, the communication time between tasks is considered as a part of the predecessor task's execution time.

The task graph model is popular for image processing in automotive systems and pedestrian detection [14]. System designers assign the criticality level of tasks based on their functionalities. However, similar to previous studies in the literature [9, 10], if a task is a predecessor of an HC task, then it is considered as an HC task. Figure 1.2 shows a real task graph (Unmanned Air Vehicle (UAV)), which is a real-life MC application task graph [9].

Since we use the utilization bound to schedule the MC tasks, the utilization of task τ_i at criticality mode l is defined as $u_i^l = \frac{WCET_i^l}{T_i}$ and $l \in \{LO, HI\}$.

2.1.1.2 Quality-of-Service (QoS)

Dropping some LC tasks in the HI mode can be used for real-time applications characterized by hard and firm deadlines. The tasks with a hard deadline can be HC tasks, and with firm deadlines can be LC tasks. The multimedia tasks are an example of firm deadlines, where skipping a video frame once in a while is better than processing it with a long delay or not processing it completely [15]. The system should execute these tasks to improve its QoS; however, the system can skip executing them in harsh situations. The QoS is defined as the percentage of executed LC tasks to all LC tasks [16, 17] ($QoS = n_L^{succ}/n_L$, where n_L is the number of all LC tasks in a task set and n_L^{succ} is the number of executed LC tasks.

2.1.1.3 System Operational Model

MC systems first start the operation in the LO mode in which all LC and HC tasks must be executed correctly before their deadlines. If the execution time of at least one HC task exceeds its low WCET ($WCET_i^{LO}$) due to unexpected conditions, the system switches to the HI mode. In this mode, since the HC tasks are supposed to execute longer, compared to the LO mode, the LC tasks are degraded to guarantee the correct execution of HC tasks before their deadlines and prevent catastrophic consequences. It stays in this mode until there is (1) no ready HC task in each core's queue and (2) ongoing HC task, executing on the processor [9, 10, 12, 18]. In the LO mode, the mapping and scheduling algorithms consider the low WCET of tasks, while in the HI mode, the algorithm schedules tasks by their high WCETs.

2.1.2 Fault-Tolerance, Fault Model, and Safety Requirements

Transient faults are the most common faults in embedded systems [17, 19, 20]. Hence, occurrence of a fault in a system is independent of the criticality levels of tasks or criticality modes of the system. Indeed, a fault occurs due to the hardware component defects, electromagnetic interference, etc. [21–23]. To tolerate transient faults, fault detection and correction mechanisms need to be applied.

For embedded safety-critical real-time systems, low-cost, low-power, and high-accuracy checker should be employed in each core. To check whether a fault occurs during the execution of a task, analogous to [13, 17], an error detection mechanism is conducted to check the correctness of the task's output at the end of the task's execution. ARGUS [24] is one of the significant checker tools to detect errors that has all the features and has been used in many recent works [25]. It can be applied to any embedded systems with less than 11% chip area overhead and also check control flow, data flow, computation, and memory access separately, at run-time. In this book, the error detection time overhead is considered in the WCET of tasks.

The task re-execution technique is one of the most popular ways to correct transient faults in embedded systems [17, 22], which we employed in this book. Some state-of-the-art works have considered that up to k transient faults may occur in one period of the application [14, 26, 27]. If the system detects a faulty task, it spends some time (μ) to discard the results of the faulty task before re-executing the task. During the design process of multi-core MC systems, we assume in Chap. 6 to tolerate up to k transient faults within a given application period, like these state-of-the-art works. However, in order to guarantee the safety requirements of an application, the required number of re-execution should be determined. Therefore, a probability factor (f_i) (Probability-of-Failure (PoF)) is considered, which indicates the probability of an unsuccessful execution of a task due to transient hardware/software faults [22]. In addition, the PFH has been exploited to measure the safety of the system. The PFH represents the rate of the average system failures in an hour [5, 8, 22]. According to safety standards, PFH estimates the failure probability of safety functions in each of the criticality levels [5, 22]. As shown in Table 2.1, five criticality levels of the exploited DO-178B safety standard, i.e., *A*, *B*, *C*, *D*, and *E*, have been illustrated, and the PFH values for all of these levels have been determined. The re-execution of the tasks is used to tolerate faults and improve the system's reliability according to Table 2.1. The number of re-executions for each criticality level of tasks, HC and LC, in each mode to guarantee safety requirement, is obtained by using the value of PFHs. We guarantee the safety requirement when designing the MC applications in Chap. 4, where we discuss later in that chapter how to calculate the number of re-executions.

2.1.3 *Hardware Architectural Modeling*

In this book, a multi-core hardware platform comprising of m cores $\{C_1, C_2, \ldots, C_m\}$ based on the ODROID XU3 board is considered. The ODROID XU3 board is DVFS-enabled and the cores can operate at multiple voltage (V) and frequency (f) levels. This board consists of two clusters with ARM cortex-A15 (big) and ARM Cortex-A7 (LITTLE) (four big cores and four LITTLE cores); hence, cores within the same cluster operate at the same *V-f* level, and also, each cluster can operate at different frequency and voltage levels. In this board, the allowed frequency is in the range of [0.2, 1.4] GHz for LITTLE cores and [0.2, 2] GHz for big cores. Besides, the voltage is in the range of [0.9, 1.3] V for LITTLE cores and [0.9, 1.3625] V for big cores.

This book focuses on employing a single-core platform based on ARM Cortex-A7 (LITTLE) cores to evaluate the approaches while designing MC application systems, as presented in Chaps. 3, 4, and 5. Besides, the multi-core platforms mentioned above are considered for MC hardware system design, as presented in Chaps. 6 and 7.

2.1.4 *Low-Power Techniques and Power Consumption Model*

Power management in electronic systems is primarily targeted toward two purposes: firstly, to minimize heat dissipation in order to improve the system's usability (for handheld devices and wearables), reliability (for safety- and mission-critical systems), etc., and secondly, the power management methods may target the minimization of the system's energy consumption. This is crucial for battery-powered and energy-harvesting systems as well as for large-scale systems. The common power management techniques used in this book are DVFS, a firmware-level technique, task re-mapping, and task scheduling, as software-level techniques. We explain these techniques and the power consumption model below.

2.1.4.1 Dynamic Voltage and Frequency Scaling (DVFS)

The total power consumption of a core is composed of static (P_s), dynamic (P_d), and independent power consumption (P_{ind}) [28, 29]. P_{ind} refers to the power related to the memory and I/O activities. DVFS technique can dynamically change the voltage and/or frequency (V-f) of one/some/all processor cores to reduce the overall system power consumption (static and dynamic), including computation, communication, and memory parts. Since the V-f level of some cores (can be named as a cluster) can be changed, it implies that all cores in a cluster must have the same V-f level. The total power consumption is given by Eq. (2.2). In this equation, I_{sub} and C_L are the subthreshold leakage current and load capacitance, respectively. In this book, we focus on decreasing P_d:

$$Pow = P_s + P_d + P_{ind} = I_{sub}V + C_L V^2 f + P_{ind} \tag{2.2}$$

in which (ρ_1 and ρ_2 are the scaling factors of frequency and voltage, respectively):

$$\begin{aligned} f_{min} &\leq f = \rho_1 \times f_{max} \leq f_{max} \\ V_{min} &\leq V = \rho_2 \times V_{max} \leq V_{max} \end{aligned} \tag{2.3}$$

Therefore, by using these scaling factors, Eq. 2.2 can be written based on the V_{max} and f_{max} as:

$$Pow = I_{sub}(\rho_2 V_{max}) + C_L(\rho_2 V_{max})^2(\rho_1 f_{max}) + P_{ind} \tag{2.4}$$

As our system is based on the ODROID XU3, some frequency levels work with the same voltage level on this board. It means, by reducing the frequency level, the voltage level does not change. Therefore, the scaling factors ρ_1 and ρ_2 do not have the same value. According to the range of frequency for big and LITTLE cores presented in Sect. 2.1.3, ρ_1 can be set in the range of [0.143, 1] for the A7 cores and [0.1, 1] for A15 cores. In addition, ρ_2 is in the range of [0.692, 1] and [0.6606, 1] for A7 and A15 cores, respectively. Although the ODROID XU3 has power sensors, they only report values for the entire cluster, not for each core. Hence, DVFS technique is employed in Chap. 7 to manage the peak power consumption and maximum temperature.

2.1.4.2 Task Re-mapping

Task re-mapping is the run-time moving of a task/application from a hot processor core to another processor core, i.e., re-map and reschedule on a colder processor core to let the hot processor core cool down. This process helps in dynamically reducing and balancing the temperature or power consumption across all processor cores in a platform [30, 31]. This technique is employed in Chap. 7 for thermal management.

2.1.4.3 Task Scheduling

Task scheduling is a process of selecting a task from an application/task set and determining where (i.e., in which core) and when to execute it [30]. Choosing a processor core from a list of available processor cores helps to reduce power consumption, especially in heterogeneous multi-core platforms. The static task scheduling process is employed with the aim of power reduction, in which the task data is known in advance. Thus, the task scheduling decision (in which processor core and the appropriate time instants to start each task's execution on the processor core) can be made at design-time to reduce power consumption. This low-power technique is employed in Chap. 6 to reduce the instantaneous power consumption and maximum temperature.

2.2 Related Works

2.2.1 Mixed-Criticality Task Scheduling Mechanisms

After proposing the MC tasks model by Vestal in 2007 [32], many research works have focused on the MC task scheduling in different operational modes and the feasibility of schedules in each mode. The majority of these papers are concerned with single-core platforms and independent tasks. The most common proposed scheduling algorithms are Earliest Deadline First with Virtual Deadline (EDF-VD) used in most papers like [4, 7, 33–37], Early Release Earliest Deadline First (ER-EDF) [38–40], and Fixed Periority (FP) [41, 42]. These algorithms have been reviewed in [43] in detail; however, the following is a brief explanation of each algorithm and the one we chose in this book.

FP scheduling algorithm is the first proposed scheduling algorithm for MC systems that is presented in [32]. In general, the scheduler gives higher priority to HC tasks and executes first the ready HC tasks of all those tasks (HC and LC tasks) at any given time, and if there is no ready HC task, LC tasks are scheduled. The algorithm can ensure that all tasks can be scheduled in LO mode, and in the case of mode switches, the scheduler can decide to drop all LC tasks or execute some of them if there is some slack time. However, some research works such as [36] studied and discussed that this FP algorithm could not handle the task scheduling while the mode is switching. As a result, some scheduling algorithms have been presented based on the Earliest Deadline First (EDF) scheduling algorithm, which can manage the mode switches, have higher utilization, and schedule more tasks in a core. ER-EDF is one of the algorithms that the complete analysis of this scheduling algorithm was presented in [39] for the first time and then extended and published in [38, 40]. In this scheduling algorithm, a maximum period (larger than the actual period) and some early release points (between actual and maximum periods) are defined for each LC task. The minimum service requirement of LC tasks is guaranteed by executing them with their maximum period. The scheduler ensures that the minimum service requirement is met in the worst-case scenario, i.e., switching the system to the HI mode. In the case of generating dynamic slack at run-time, the early release points are used for LC tasks to provide opportunities for them to release more frequently and improve their QoS. As can be realized, in this algorithm, the quality of output results for LC tasks might be reduced in the LO mode, which is not acceptable in some applications.

EDF-VD is the most common scheduling algorithm in MC system in the last decade. The complete analysis of the EDF-VD scheduling algorithm was first presented in [44]. In this scheduling algorithm, a virtual deadline is defined for each HC task, which is obtained by multiplying the actual deadline by x $(0 < x < 1)$. This policy leads to providing a higher priority for HC tasks in the scheduling algorithm. When the system is in the LO mode, the virtual deadlines are used for HC tasks in the EDF scheduler, and also all of the HC and LC tasks are executed before their deadlines. Nevertheless, when the system switches to the HI mode, the actual deadlines of HC tasks are used in the EDF scheduler and all/some of the LC tasks are dropped. An appropriate interval of x and the required conditions for the EDF-VD algorithm for scheduling a given set of MC tasks in each operational mode are presented in detail in [4]. In this book, we apply the EDF-VD algorithm to schedule independent MC tasks in a single processor and present the required and sufficient conditions in each chapter according to the employed task execution policy. Since in Chap. 4, a new task parameter is defined to schedule more LC tasks in the system, the optimum value of x and the required and sufficient conditions are presented based on the new parameter.

2.2.2 *QoS Improvement Methods in Mixed-Criticality Systems*

In most safety-critical applications, dropping LC tasks causes serious service interruptions for those LC tasks. Therefore, some approaches have been presented to execute more LC tasks and improve the QoS. We can divide these approaches in two categories: (1) approaches that improve QoS through appropriate WCET adjustment for HC tasks and (2) approaches that improve QoS through task dropping analysis in the HI mode. Below is an explanation of each category.

2.2.2.1 QoS Improvement Through WCET Adjustment

A significant number of papers have been published in the last decade regarding the design of MC systems. Burns and Davis [43] provided a comprehensive study in this field; however, since our focus is on improving the timing behavior and QoS improvement of these MC systems and WCET analysis at both design-time and run-time, we mostly focus on the works presented for designing these systems with similar scope. Table 2.2 summarizes these works.

Table 2.2 A brief overview of the state-of-the-art MC approaches in QoS improvement through WCET adjustment

#	Related works	Dynamic QoS-aware	Low WCET adjustment	Design-/run-time	Mode switching prob. determination
1	[4, 35]	×	✓	✓/×	×
2	[41, 50]	×	✓	✓/×	×
3	[51, 52]	×	✓	×/✓	×
4	[53]	×	✓	×/✓	✓
5	[38, 40]	✓	×	×/✓	×

The MC task model has been presented by Vestal in [32] for the first time and introduced different WCET levels for tasks. However, the author has not discussed how these WCETs are obtained and how often the system switches to the HI mode based on the design. The authors have discussed a bit further in [45] how different WCETs can be defined and determined. As an example, they can be determined at different levels of accuracy with different degrees of confidence by limiting the programming constructs, used in implementing the task. However, this approach does not involve any analysis. Most of the approaches, such as [4, 7, 35–37, 46–49], generally count the low WCETs as a percentage of the high WCETs to be employed in system design, and then these values are not changing during the run-time (shown some of them in row 1 of Table 2.2). This policy may waste the system utilization, or cause frequent mode switches at run-time, which disturbs the LC tasks and reduces their QoS. Although the efficiency of these approaches has been evaluated for different percentages of high WCET, there is no scalable approach for determining the WCETs for all criticality levels. In addition, these estimations are not accurate since WCET and Actual Execution Times (AETs) do not always have a linear relationship and therefore cannot be employed for dynamic QoS improvement.

Besides, a few studies such as [41, 50], have focused on probability distributions in MC systems by exploiting Extreme Value Theory (EVT) [54] for timing analysis, which are presented in row 2 in Table 2.2. Note that EVT is a branch of statistics which estimates and models the probability distribution of extreme events. In the field of real-time systems, EVT is exploited to determine WCETs. Applying these estimation methods has some open challenges, such as the required number of execution times for a sample and its incomplete representativity identification and evaluation that make it uncertain and unreliable [55–57]. Researchers in [57] have recently exploited this probabilistic information and proposed a technique to optimize the energy consumption of MC systems by finding the optimum core speed in the LO mode and based on that, obtaining the low WCET. However, their system operation model definition for running the LC tasks is different from the popular MC model. In this system, all LC and HC tasks are executed in both LO mode and HI mode, and the authors have obtained the low WCET for HC tasks to investigate the trade-off between the minimum core frequency (that leads to energy minimization) and probability of mode switching. Switching the system to the HI mode causes an increase in processor frequency to guarantee all task schedulability before their deadlines. In fact, although this method improves energy consumption, it causes to schedule fewer LC tasks in the system which leads to underutilization.

A few studies [51, 52] have determined the low WCET of tasks at run-time (row 3), based on their overall processing requirements and actual execution times. However, there is no guarantee at design-time on optimal use of the system utilization and LC tasks' execution. From the mode switching probability perspective, some research works such as [53] have addressed mode switching probability in MC systems and how to have the safe mode switching at run-time. However, the relation between the HC tasks' low WCET and mode switching probability has not been discussed. Besides, the goal of these methods in rows 3 and 4 is to postpone the mode switches for a long time while only guaranteeing a minimum QoS for LC tasks.

Some research works such as [38, 40], shown in row 5, have considered the low WCET as a percentage of high WCET and improved the QoS at run-time by exploiting the accumulated dynamic slack generated by early completion of HC tasks. Since the dynamic slack is considered as a wrapper task that has a deadline [40, 58] and cannot be used anytime, these approaches do not use the system utilization optimally to improve the QoS.

Therefore, an appropriate low WCET analysis of MC tasks at both design- and run-time is needed to reduce the use of WCET estimation methods and improve the confidence in the WCET's values, service adaptation, and processor utilization [43].

2.2.2.2 QoS Improvement Through Task Dropping Analysis

There are some existing studies in the context of MC systems, which have focused on proposing approaches for managing different aspects of the MC system design in terms of task schedulability and QoS improvement while guaranteeing the safety requirements at design-time. A few efforts have also been conducted to manage these parameters at run-time. Although [43] gives a comprehensive study in the field of QoS improvement in MC systems in the run- and design-time phases, this sub-section provides an overview of the existing studies in QoS improvement through task dropping analysis. Table 2.3 summarizes these works with their respective properties, like run-/design-time approach, and QoS improvement (offline/online manner), safety requirement consideration while dropping LC tasks, and the policy of LC task dropping in the HI mode.

Most of the existing approaches design MC systems by dropping all LC tasks when the system switches to the HI mode in order to guarantee the correct execution of HC tasks. Although these approaches may reduce the safety requirement of tasks, dropping all LC tasks causes serious service interruptions for those LC tasks. Since most MC systems are safety-related and real time, the task schedulability in terms of QoS improvement is typically analyzed at design-time to guarantee the correct execution of tasks before their deadlines to prevent catastrophic damages while the

Table 2.3 A brief overview of the state-of-the-art studies in QoS improvement through task dropping analysis

#	Related works	Safety requirement guarantee	Design-/Run-time	offline/online QoS opt.	LC task drop policy
1	[35, 47, 59–62]	×	✓/×	✓/×	Drop freq.
2	[36, 42, 63, 64]	×	✓/×	✓/×	Degradation
3	[5, 8]	✓	✓/×	✓/×	Degradation or drop all
4	[46, 65]	✓	✓/×	✓/×	Drop freq.
5	[66]	×	×/✓	×/×	Drop all
6	[37–39, 67, 68]	×	×/✓	×/✓	Degradation
7	[69–72]	×	×/✓	×/✓	Drop freq.

system is operating. Row 1 of Table 2.3 shows the research works that have focused on MC system design to improve the LC tasks' QoS in the worst-case scenario in the case of mode switches to the HI mode. In these approaches, executing the minimum number of instances (i.e., dropping fewer instances of LC tasks) in the HI mode is ensured. However, they do not guarantee (1) to not drop LC tasks frequently and (2) safety requirements. Besides, this is despite the fact that the system does not operate in the worst-case scenario at run-time in most cases.

In addition, recent studies have provided techniques to improve the minimum service level of LC tasks in the HI mode (instead of dropping all LC tasks) by reducing the WCET of LC tasks in the HI mode or increasing their period in the HI mode (presented in row 2 of Table 2.3). Indeed, they degrade the service level of LC tasks that the minimum service level would be guaranteed by their techniques while not guaranteeing the service requirements. However, the common part of all previous methods is their consideration on an MC model in which LC tasks are dropped or degraded when the system switches to the HI mode. Thus, none of these algorithms can be applied to MC tasks that LC tasks could not be frequently dropped or postponed for a long time.

A few papers [46, 65] have improved QoS of LC tasks in the HI mode while guaranteeing the tasks' safety requirements (rows 3 and 4). Although these research works have tried to increase QoS of LC tasks in the HI mode and guarantee the safety requirements, they may drop/degrade LC tasks in the HI mode frequently, which is not acceptable, especially for mission-critical tasks. Degrading or dropping LC tasks in a frequent manner without any restrictions in the HI mode is not desirable and may negatively affect the safety and even lead to catastrophic consequences.

From the MC task scheduling perspective at run-time, Sigrist et al. [66] (row 5) have studied the recent task scheduling mechanisms and evaluated the effect of run-time overheads, such as task execution monitoring, overrunning detection, and mode switching. However, they have not improved the task scheduling and QoS at run-time.

To improve the LC tasks' QoS at run-time, some state-of-the-art works have presented the run-time adaptability mechanisms by exploiting the accumulated dynamic slack to execute more LC tasks in the HI mode through fewer LC task dropping/degradation (shown in rows 6 and 7). In [69], a run-time schedulability analysis has been presented, and LC tasks are executed in free slack time if the conditions are met. Researchers in [71] have also presented an effective solution for reducing the number of mode switches and consequently LC task dropping by using the generated dynamic slack for executing HC tasks when they overrun. Bate et al. [72] have also proposed a protocol that handles mode switches and ensures that LC tasks are executed more frequently. However, in such three works, some LC tasks may be dropped frequently and continuously when the system switches to the HI mode, which is unacceptable in any situation in some MC systems. In addition, all dynamic slack is not exploited in [70] properly since the algorithm only uses the dynamic slack generated by HC tasks' execution.

The downsides of the previous approaches have motivated us to study the MC systems' behavior at both design- and run-time and propose methods to improve the LC tasks' QoS by task dropping analysis and not let them drop frequently.

2.2.3 QoS-Aware Power and Thermal Management in Multi-core Mixed-Criticality Systems

In this subsection, since we also focus on QoS-aware multi-core MC system design while managing power consumption, we overview the literature works in the following, which have considered some or all of our target optimization objectives or have used the same task and system model. Many previous works in the context of MC systems with dependent task model have just focused on proposing techniques to show how to efficiently map and schedule dependent tasks in both design- and run-time phases. Since our focus is on QoS-aware MC task scheduling to manage power and temperature, we only consider the works presented for MC or non-MC systems with similar scope. Generally, the related works on power and thermal management for real-time systems can be classified based on the assumed system model, like MC or non-MC systems and dependent or independent tasks, and also the target optimization objectives of QoS, peak power, average power, or maximum temperature. Table 2.4 summarizes the recent works with different target optimization objectives and assumed task models along with used low-power techniques and whether each approach has considered the timing overheads of the scheduler or changing *V-f* levels.

There are some algorithms presented for independent tasks such as EDF-VD used in [4], or using different scheduling policies for different criticality levels [6]. Hence, these algorithms are presented for independent periodic tasks and cannot be applied to tasks with precedence constraints. Besides, as can be seen in row 1 of Table 2.4, some papers have considered periodic MC tasks with data dependency but none of them have considered power management. These papers have focused on the feasibility of schedules and meeting the timing constraints. From the perspective

Table 2.4 Summary of state-of-the-art approaches in power-aware MC system hardware design

#	Related works	MC tasks	DAG model	QoS	Peak power	Avg. power	Temp.	Low-power technique	Timing overhead
1	[9, 12, 29, 62, 76, 77]	✓	✓	✓	×	×	×	×	×
2	[19, 78–82]	✓	×	×	×	✓	×	DVFS	×
3	[83]	✓	×	×	×	✓	✓	DVFS	×
4	[84, 85]	×	×	×	✓	×	×	Task sch.	×
5	[86]	×	✓	×	✓	×	×	Task sch.	×
6	[87]	×	✓	×	✓	✓	×	DVFS	×
7	[88–93]	×	✓	×	×	✓	×	DVFS	×
8	[94]	×	✓	×	×	✓	×	DVFS	✓
9	[95–99]	×	✓	×	×	✓	✓	DVFS	×
10	[100]	×	✓	×	×	✓	✓	DVFS	✓

of guaranteeing LC tasks' minimum service level, most existing MC scheduling algorithms like what presented in [10, 73–75] discard or degrade LC tasks when the system switches to the HI mode. It causes serious service interruption for LC tasks. Therefore, in addition to power management in MC systems, improving the QoS of LC tasks would be significant. There are few research works, like those presented in row 1, that have improved the QoS while scheduling dependent MC tasks. On the other hand, in [9, 12], the dependent MC tasks are scheduled in multi-core systems with consideration of fault occurrence possibilities, but with no power or thermal management.

From the perspective of power management in MC systems, some works such as what are presented in row 2 have presented methods to minimize the average power consumption in MC systems theoretically in which systems are single or multi-core and tasks are independent. In general, they only optimize the average power in the LO mode in simulation by using DVFS technique. When the system switches to the HI mode, all HC tasks are executed with the highest frequency; and all LC tasks are dropped. Indeed, they interrupt the minimum service level of LC tasks in the HI mode. As a result, in the HI mode, with higher frequency, the peak power consumption of the system may increase significantly, which is not admissible.

Furthermore, there is a paper [83] that has considered thermal management in MC systems (third row of Table 2.4). The researchers minimize the temperature of single-core processors by finding the optimum speed for each task in the design-time phase. Hence, they discard LC tasks when the system switches to the HI mode, which is not acceptable in many MC applications. Besides, they do not consider the latency of changing the V-f level at run-time, which may cause deadline violation and, consequently, catastrophic consequences.

Some studies concentrate on peak power management in multi-core systems at design-time (rows 4–6) by using DVFS or task scheduling techniques. These papers have only considered hard real-time tasks with one criticality level which is not practical for MC. It should be mentioned that authors in [86] work on the dependent task model in which the execution of some tasks is postponed to manage the simultaneous peak power consumption. It is not suitable for MC tasks, especially in the HI mode.

The previous works in the context of power or thermal management in non-MC systems that use DVFS by considering the dependent task model are shown in rows (7–10) of Table 2.4.

Researchers in [88–93] (shown in row 7) have used slack reclamation to apply online DVFS to the system while executing dependent tasks. Kang et al. [89] propose an algorithm that uses dynamic voltage scaling to minimize energy without considering the tasks' deadlines, which is not suitable for MC systems. Researchers in [90–92] suggest a run-time energy management technique that uses reclaimable slack for the immediately ready task to decrease average power. Their results show that the power can be reduced; however, the possibilities of looking further ahead into the future execution of the following tasks to have better results are not explored. Besides, in [93], the authors have considered two types of tasks, best effort and real-time, and they have just used the dynamic slack for the next real-time task to reduce its *V-f* level, which is inefficient. There is an aggressive slack reclamation algorithm, presented by [88], in which the generated dynamic slack is checked to be able to be used for the next task if the remaining tasks could complete their execution before the deadline. However, in general, the average energy consumption is reduced, but this algorithm has focused more on meeting the deadlines, while we target both energy minimization and meeting the deadlines.

In addition, from the DVFS latency perspective, some few works, like what has been presented in [94], have presented a method to minimize energy in a multi-core platform by using the DVFS technique. Researchers in [94] have considered a task graph model running on the cluster-based platform. They have also considered the latency of changing frequency in their paper. As shown in row 8, they have not considered peak power or thermal management, and also, their method is not suitable for MC systems where tasks have different criticality levels.

Row 9 demonstrates the works which focus on average power and maximum temperature reduction in a system with dependent non-MC tasks with no timing overhead consideration. As an example, in [97], a look-up table for each task is generated in the offline phase, which contains the optimum voltage and frequency settings for each core for every possible run-time condition, task execution time, and core temperature measurement. The memory overhead incurred in generating these tables may not be desirable, especially for multi-core systems with many tasks and cores. Timing overhead of changing *V-f* levels has been considered in [100] while reducing the average power consumption and maximum temperature (shown in row 10).

As a result, more works are needed to be studied to reduce the peak power consumption and maximum temperature in MC systems while improving the QoS and considering the timing overheads of changing *V-f* level and task scheduler in order to guarantee the correct execution of tasks before their deadlines.

2.3 Conclusions

In this chapter, we first discussed the model and assumption that are used in this book. We introduced the general MC task model and system operational model and defined the QoS metric. Then, we provide a short explanation of the fault model, hardware architectural model, power model, and low-power techniques that are used in this book.

Available MC scheduling algorithms were discussed, and the chosen scheduling algorithm (EDF-VD) to be used in this book was highlighted. We summarized the related literature works in QoS-aware MC application design and QoS-aware MC hardware design, where power management is one of the objectives while designing such systems. We showed the works with different target objectives in tables for better understanding.

According to our discussion in Sect. 2.2, some objectives are needed to be considered and improved in MC system design. In the following chapters, we present the contributions of this book and compare them to some relevant state-of-the-art works.

References

1. R. Ernst and M. Di Natale. "Mixed Criticality Systems—A History of Misconceptions?" In: *IEEE Design & Test* 33.5 (2016), pp. 65–74.
2. Michael Zimmer et al. "FlexPRET: A processor platform for mixed-criticality systems". In: *Proc. of Real-Time and Embedded Technology and Applications Symposium (RTAS)*. IEEE. 2014, pp. 101–110.
3. Leslie A Johnson. "DO-178B, Software considerations in airborne systems and equipment certification". In: *Crosstalk, October* 199 (1998), pp. 11–20.
4. S. Baruah et al. "The preemptive uniprocessor scheduling of mixed-criticality implicit-deadline sporadic task systems". In: *Proc. of Euromicro Conference on Real-Time Systems (ECRTS)*. 2012, pp. 145–154.
5. P. Huang, H. Yang, and L. Thiele. "On the scheduling of fault-tolerant mixed-criticality systems". In: *Proc. on Design Automation Conference (DAC)*. 2014, pp. 1–6.
6. James H Anderson, Sanjoy K Baruah, and Björn B Brandenburg. "Multicore operating-system support for mixed criticality". In: *Proc. of the Workshop on Mixed Criticality: Roadmap to Evolving UAV Certification*. Vol. 4. Citeseer. 2009, p. 7.
7. Sanjoy Baruah et al. "Preemptive Uniprocessor Scheduling of Mixed-Criticality Sporadic Task Systems". In: *Journal of the ACM (JACM)* 62.2 (2015), 14:1–14:33.
8. Luyuan Zeng, Pengcheng Huang, and Lothar Thiele. "Towards the Design of Fault-tolerant Mixed-criticality Systems on Multicores". In: *Proc. of Compilers, Architectures and Synthesis for Embedded Systems (CASES)*. Pittsburgh, Pennsylvania, 2016, 6:1–6:10. ISBN: 978-1-4503-4482-1.
9. R. Medina, E. Borde, and L. Pautet. "Availability enhancement and analysis for mixed-criticality systems on multi-core". In: *Proc. on Design, Automation & Test in Europe Conference & Exhibition (DATE)*. 2018, pp. 1271–1276.
10. Sanjoy Baruah. "The federated scheduling of systems of mixed-criticality sporadic DAG tasks". In: *Proc. of IEEE Real-Time Systems Symposium (RTSS)*. 2016, pp. 227–236.

11. Mostafa Jafari-Nodoushan, Bardia Safaei, and Alireza Ejlali. "Leakage-Aware Battery Lifetime Analysis Using the Calculus of Variations". In: *IEEE Transactions on Circuits and Systems I: Regular Papers* (2020).
12. Roberto Medina, Etienne Borde, and Laurent Pautet. "Scheduling Multi-periodic Mixed-Criticality DAGs on Multi-core Architectures". In: *Proc. of IEEE Real-Time Systems Symposium (RTSS)*. 2018, pp. 254–264.
13. Mohammad Salehi, Alireza Ejlali, and Bashir M Al-Hashimi. "Two-phase low-energy N-modular redundancy for hard real-time multi-core systems". In: *IEEE Transactions on Parallel and Distributed Systems (TPDS)* 27.5 (2016), pp. 1497–1510.
14. Viacheslav Izosimov, Petru Eles, and Zebo Peng. "Value-based scheduling of distributed fault-tolerant real-time systems with soft and hard timing constraints". In: *Proc. of the IEEE Workshop on Embedded systems for real-time multimedia (ESTIMedia)*. 2010, pp. 31–40.
15. Giorgio C Buttazzo. *Hard real-time computing systems: predictable scheduling algorithms and applications*. Vol. 24. Springer Science & Business Media, 2011.
16. Zheng Li and Shuibing He. "Fixed-Priority Scheduling for Two-Phase Mixed-Criticality Systems". In: *ACM Transactions on Embedded Computing Systems (TECS)* 17.2 (2018), pp. 1–20.
17. B. Ranjbar et al. "FANTOM: Fault Tolerant Task-Drop Aware Scheduling for Mixed-Criticality Systems". In: *IEEE Access* 8 (2020), pp. 187232–187248. DOI: 10.1109/ACCESS.2020.3031039.
18. B. Ranjbar et al. "Online Peak Power and Maximum Temperature Management in Multi-core Mixed-Criticality Embedded Systems". In: *Proc. of Euromicro Conference on Digital System Design (DSD)*. 2019, pp. 546–553.
19. Zheng Li et al. "Empirical study of energy minimization issues for mixed-criticality systems with reliability constraint". In: *Proc. 1st Workshop on Low-Power Dependable Computing (LPDC)*. 2014, pp. 3–5.
20. Siva Satyendra Sahoo, Behnaz Ranjbar, and Akash Kumar. "Reliability-Aware Resource Management in Multi-/Many-Core Systems: A Perspective Paper". In: *Journal of Low Power Electronics and Applications* 11.1 (2021), p. 7.
21. Jin Jiang and Xiang Yu. "Fault-tolerant control systems: A comparative study between active and passive approaches". In: *Annual Reviews in control* 36.1 (2012), pp. 60–72.
22. Israel Koren and C Mani Krishna. *Fault-tolerant systems*. Morgan Kauf-mann,2020.
23. Elena Dubrova. "Fundamentals of dependability". In: *Fault-Tolerant Design*. Springer, 2013, pp. 5–20.
24. A. Meixner, M. E. Bauer, and D. Sorin. "Argus: Low-Cost, Comprehensive Error Detection in Simple Cores". In: *Proc. of IEEE/ACM International Symposium on Microarchitecture (MICRO)*. 2007, pp. 210–222. DOI: 10.1109/MICRO.2007.18.
25. Jon Perez Cerrolaza et al. "Multi-Core Devices for Safety-Critical Systems: A Survey". In: *ACM Computing Surveys (CSUR)* 53.4 (2020), pp. 1–38.
26. V. Izosimov et al. "Scheduling of Fault-Tolerant Embedded Systems with Soft and Hard Timing Constraints". In: *Proc. on Design, Automation & Test in Europe Conference & Exhibition (DATE)*. 2008, pp. 915–920.
27. M. Salehi et al. "Two-State Checkpointing for Energy-Efficient Fault Tolerance in Hard Real-Time Systems". In: *IEEE Transactions on Very Large Scale Integration (VLSI) Systems* 24.7 (2016), pp. 2426–2437.
28. Amir Taherin, Mohammad Salehi, and Alireza Ejlali. "Stretch: exploiting service level degradation for energy management in mixed-criticality systems". In: *Proc. of CSI Symposium on Real-Time and Embedded Systems and Technologies (RTEST)*. IEEE. 2015, pp. 1–8.
29. J. Li et al. "Mixed-Criticality Federated Scheduling for Parallel Real-Time Tasks". In: *Proc. of IEEE Real-Time and Embedded Technology and Applications Symposium (RTAS)*. 2016, pp. 1–12.
30. Saad Zia Sheikh and Muhammad Adeel Pasha. "Energy-Efficient Multicore Scheduling for Hard Real-Time Systems: A Survey". In: *ACM Trans. Embed. Comput. Syst. (TECS)* 17.6 (2018). DOI: 10.1145/3291387.

31. Jörg Henkel and Nikil Dutt. *Dependable embedded systems*. Springer Nature, 2021.
32. Steve Vestal. "Preemptive scheduling of multi-criticality systems with varying degrees of execution time assurance". In: *Proc. of Real-Time Systems Symposium (RTSS)*. IEEE. 2007, pp. 239–243.
33. Sanjoy Baruah et al. "Scheduling real-time mixed-criticality jobs". In: *IEEE Transactions on Computers (TC)* 61.8 (2012), pp. 1140–1152.
34. Sanjoy Baruah, Haohan Li, and Leen Stougie. "Towards the design of certifiable mixed-criticality systems". In: *Proc. of Real-Time and Embedded Technology and Applications Symposium (RTAS)*. IEEE. 2010, pp. 13–22.
35. D. Liu et al. "Scheduling Analysis of Imprecise Mixed-Criticality Real-Time Tasks". In: *IEEE Transactions on Computers (TC)* 67.7 (2018), pp. 975–991.
36. D. Liu et al. "EDF-VD Scheduling of Mixed-Criticality Systems with Degraded Quality Guarantees". In: *Proc. on IEEE Real-Time Systems Symposium (RTSS)*. 2016, pp. 35–46.
37. G. Chen et al. "Utilization-Based Scheduling of Flexible Mixed-Criticality Real-Time Tasks". In: *IEEE Transactions on Computers (TC)* 67.4 (2018), pp. 543–558.
38. Hang Su and Dakai Zhu. "An elastic mixed-criticality task model and its scheduling algorithm". In: *Proc. on Design, Automation & Test in Europe Conference & Exhibition (DATE)*. IEEE. 2013, pp. 147–152.
39. Hang Su, Dakai Zhu, and Daniel Mossé. "Scheduling algorithms for elastic mixed-criticality tasks in multicore systems". In: *Proc. on Embedded and Real-Time Computing Systems and Applications (RTCSA)*. 2013, pp. 352–357.
40. H. Su, D. Zhu, and S. Brandt. "An Elastic Mixed-Criticality Task Model and Early-Release EDF Scheduling Algorithms". In: *ACM Trans. Des. Autom. Electron. Syst. (TODAES)* 22.2 (2016), pp. 1–25.
41. Dorin Maxim et al. "Probabilistic Analysis for Mixed Criticality Systems Using Fixed Priority Preemptive Scheduling". In: *Proc. of Euromicro Conference on Real-Time Systems (RTNS)*. 2017, pp. 237–246.
42. Alan Burns and Sanjoy Baruah. "Towards a more practical model for mixed criticality systems". In: *Workshop on Mixed-Criticality Systems*. 2013.
43. Alan Burns and Robert I. Davis. "A Survey of Research into Mixed Criticality Systems". In: *ACM Computing Surveys (CSUR)* 50.6 (2017), pp. 1–37.
44. Sanjoy K Baruah et al. "Mixed-criticality scheduling of sporadic task systems". In: *European Symposium on Algorithms*. 2011, pp. 555–566.
45. S. Baruah and S. Vestal. "Schedulability Analysis of Sporadic Tasks with Multiple Criticality Specifications". In: *Proc. of Euromicro Conference on Real-Time Systems (ECRTS)*. 2008, pp. 147–155. DOI: 10.1109/ECRTS.2008.26.
46. Zaid Al-bayati et al. "A four-mode model for efficient fault-tolerant mixed-criticality systems". In: *Proc. on Design, Automation & Test in Europe Conference & Exhibition (DATE)*. 2016, pp. 97–102.
47. Z. Guo et al. "Uniprocessor Mixed-Criticality Scheduling with Graceful Degradation by Completion Rate". In: *Proc. on IEEE Real-Time Systems Symposium (RTSS)*. 2018, pp. 373–383.
48. Chuancai Gu et al. "Partitioned mixed-criticality scheduling on multi-processor platforms". In: *Proc. on Design, Automation & Test in Europe Conference & Exhibition (DATE)*. 2014, p. 292.
49. Z. Guo, L. Santinelli, and K. Yang. "EDF Schedulability Analysis on Mixed-Criticality Systems with Permitted Failure Probability". In: *Proc. of the International Conference on Embedded and Real-Time Computing Systems and Applications (RTCSA)*. 2015, pp. 187–196.
50. Luca Santinelli and Laurent George. "Probabilities and mixed-criticalities: the probabilistic c-space". In: *Proc. on IEEE Real-Time Systems Symposium (RTSS)*. 2015.
51. X. Gu and A. Easwaran. "Dynamic Budget Management with Service Guarantees for Mixed-Criticality Systems". In: *Proc. on IEEE Real-Time Systems Symposium (RTSS)*. 2016, pp. 47–56.
52. X. Gu and A. Easwaran. "Dynamic budget management and budget reclamation for mixed-criticality systems". In: *Real-Time Systems* 55.3 (2019), pp. 552–597.

53. B. Hu et al. "FFOB: Efficient online mode-switch procrastination in mixed-criticality systems". In: *Real-Time Systems* 55.3 (2019), pp. 471–513.
54. Laurens De Haan and Ana Ferreira. *Extreme value theory: an introduction*. Springer Science & Business Media, 2007.
55. S. Jiménez Gil et al. "Open Challenges for Probabilistic Measurement-Based Worst-Case Execution Time". In: *IEEE Embedded Systems Letters* 9.3 (2017), pp. 69–72.
56. Federico Reghenzani, Luca Santinelli, and William Fornaciari. "Dealing with Uncertainty in PWCET Estimations". In: *ACM Trans. Embed. Comput. Syst. (TECS)* 19.5 (Sept. 2020).
57. Ashikahmed Bhuiyan et al. "Optimizing Energy in Non-Preemptive Mixed-Criticality Scheduling by Exploiting Probabilistic Information". In: *IEEE Transactions on Computer-Aided Design of Integrated Circuits and Systems (TCAD)* 39.11 (2020), pp. 3906–3917. DOI: 10.1109/TCAD.2020.3012231.
58. Dakai Zhu and Hakan Aydin. "Reliability-Aware Energy Management for Periodic Real-Time Tasks". In: *IEEE Transactions on Computers* 58.10 (2009), pp. 1382–1397. DOI: 10.1109/TC.2009.56.
59. Oliver Gettings, Sophie Quinton, and Robert I. Davis. "Mixed Criticality Systems with Weakly-Hard Constraints". In: *Proc. of International Conference on Real Time and Networks Systems (RTNS)*. Lille, France: Association for Computing Machinery, 2015, pp. 237–246. ISBN: 9781450335911.
60. Saravanan Ramanathan and Arvind Easwaran. "Mixed-criticality scheduling on multiprocessors with service guarantees". In: *Proc. of IEEE International Symposium on Real-Time Distributed Computing (ISORC)*. 2018, pp. 17–24.
61. Risat Mahmud Pathan. "Improving the quality-of-service for scheduling mixed-criticality systems on multiprocessors". In: *Proc. of Euromicro Conference on Real-Time Systems (ECRTS)*. 2017.
62. Risat Mahmud Pathan. "Improving the Schedulability and Quality of Service for Federated Scheduling of Parallel Mixed-Criticality Tasks on Multiprocessors". In: *Proc. of Euromicro Conference on Real-Time Systems (ECRTS)*. Vol. 106. 2018, 12:1–12:22.
63. K. Yang and Z. Guo. "EDF-Based Mixed-Criticality Scheduling with Graceful Degradation by Bounded Lateness". In: *Proc. of Embedded and Real-Time Computing Systems and App. (RTCSA)*. 2019, pp. 1–6.
64. Lin Huang et al. "Improving QoS for global dual-criticality scheduling on multiprocessors". In: *Proc. of IEEE International Conference on Embedded and Real-Time Computing Systems and Applications (RTCSA)*. IEEE. 2019, pp. 1–11.
65. Jonah Caplan et al. "Mapping and scheduling mixed-criticality systems with on-demand redundancy". In: *IEEE Transaction on Computers* 67.4 (2018), pp. 582–588.
66. L. Sigrist et al. "Mixed-criticality runtime mechanisms and evaluation on multicores". In: *Proc. IEEE Real-Time and Embedded Technology and Applications Symposium (RTAS)*. 2015, pp. 194–206.
67. Pengcheng Huang et al. "Run and be safe: Mixed-criticality scheduling with temporary processor speedup". In: *Proc. on Design, Automation & Test in Europe Conference & Exhibition (DATE)*. IEEE. 2015, pp. 1329–1334.
68. J. Boudjadar et al. "Combining Task-level and System-level Scheduling Modes for Mixed Criticality Systems". In: *Proc. of International Symposium on Distributed Simulation and Real Time Applications (DS-RT)*. 2019, pp. 1–10.
69. Jaewoo Lee et al. "MC-ADAPT: Adaptive Task Dropping in Mixed-Criticality Scheduling". In: *ACM Transactions on Embedded Computing Systems (TECS)* 16.5s (2017).
70. Z. Li, S. Ren, and G. Quan. "Dynamic Reservation-Based Mixed-Criticality Task Set Scheduling". In: *Proc. of IEEE Intl. Conf. on High Performance Computing and Communications, IEEE Intl. Symp. on Cyberspace Safety and Security, IEEE Intl. Conf. on Embedded Software and Syst (HPCC,CSS, ICESS)*. 2014, pp. 603–610.
71. Biao Hu et al. "On-the-fly fast overrun budgeting for mixed-criticality systems". In: *Proc. of International Conference on Embedded Software (EMSOFT)*. 2016, pp. 1–10. DOI: 10.1145/2968478.2968491.

72. Iain Bate, Alan Burns, and Robert I. Davis. "A Bailout Protocol for Mixed Criticality Systems". In: *Proc. of Euromicro Conference on Real-Time Systems (ECRTS)*. 2015, pp. 259–268. DOI: 10.1109/ECRTS.2015.30.
73. Dario Socci et al. "Multiprocessor scheduling of precedence-constrained mixed-critical jobs". In: *Proc. of International Symposium on Real-Time Distributed Computing (ISORC)*. 2015, pp. 198–207.
74. D. Socci et al. "Priority-based scheduling of mixed-critical jobs". In: *Real-Time Systems* 55.4 (2019), pp. 709–773.
75. Roberto Medina, Etienne Borde, and Laurent Pautet. "Directed acyclic graph scheduling for mixed-criticality systems". In: *Ada-Europe International Conference on Reliable Software Technologies*. Springer. 2017, pp. 217–232.
76. C. Bolchini and A. Miele. "Reliability-Driven System-Level Synthesis for Mixed-Critical Embedded Systems". In: *IEEE Transactions on Computers (TC)* 62.12 (2013), pp. 2489–2502.
77. Junchul Choi, Hoeseok Yang, and Soonhoi Ha. "Optimization of Fault-Tolerant Mixed-Criticality Multi-Core Systems with Enhanced WCRT Analysis". In: *ACM Trans. Des. Autom. Electron. Syst. (TODAES)* 24.1 (2018), pp. 1–26.
78. Pengcheng Huang et al. "Energy efficient dvfs scheduling for mixed-criticality systems". In: *Proc. on Embedded Software (EMSOFT)*. 2014, pp. 1–10.
79. Ijaz Ali, Jun-ho Seo, and Kyong Hoon Kim. "A dynamic power-aware scheduling of mixed-criticality real-time systems". In: *Proc. on Computer and Information Technology; Ubiquitous Computing and Communications; Dependable, Autonomic and Secure Computing; Pervasive Intelligence and Computing (CIT/IUCC/DASC/PICOM)*. 2015, pp. 438–445.
80. Sujay Narayana et al. "Exploring energy saving for mixed-criticality systems on multi-cores". In: *Proc. on Real-Time and Embedded Technology and Applications Symposium (RTAS)*. 2016, pp. 1–12.
81. A. Taherin, M. Salehi, and A. Ejlali. "Reliability-Aware Energy Management in Mixed-Criticality Systems". In: *IEEE Transactions on Sustainable Computing (TSUSC)* 3.3 (2018), pp. 195–208.
82. M. A. Awan, D. Masson, and E. Tovar. "Energy efficient mapping of mixed criticality applications on unrelated heterogeneous multicore platforms". In: *Proc. on IEEE Symposium on Industrial Embedded Systems (SIES)*. 2016, pp. 1–10.
83. Tiantian Li et al. "TA-MCF: Thermal-Aware Fluid Scheduling for Mixed-Criticality System". In: *Journal of Circuits, Systems and Computers (JCSC)* 28.02 (2019), p. 1950029.
84. Jinkyu Lee, Buyoung Yun, and Kang G Shin. "Reducing peak power consumption inmulti-core systems without violating real-time constraints". In: *IEEE Transactions on Parallel and Distributed Systems (TPDS)* 25.4 (2014), pp. 1024–1033.
85. Waqaas Munawar et al. "Peak Power Management for scheduling real-time tasks on heterogeneous many-core systems". In: *Proc. of the International Conference on Parallel and Distributed Systems (ICPADS)*. 2014, pp. 200–209.
86. BongKi Lee et al. "Peak power reduction methodology for multi-core systems". In: *Proc. of the International SoC Design Conference (ISOCC)*. 2010, pp. 233–235.
87. M. Ansari et al. "Peak power management to meet thermal design power in fault-tolerant embedded systems". In: *IEEE Transactions on Parallel and Distributed Systems (TPDS)* 30.1 (2019), pp. 161–173.
88. Jian-Jia Chen, Chuan-Yue Yang, and Tei-Wei Kuo. "Slack reclamation for real-time task scheduling over dynamic voltage scaling multiprocessors". In: *Proc. on Sensor Networks, Ubiquitous, and Trustworthy Computing (SUTC)*. Vol. 1. 2006, pp. 1–8.
89. Jaeyeon Kang and Sanjay Ranka. "Dynamic slack allocation algorithms for energy minimization on parallel machines". In: *Journal of Parallel and Distributed Computing (JPDC)* 70.5 (2010), pp. 417–430.
90. D. Zhu, R. Melhem, and B. R. Childers. "Scheduling with dynamic voltage/speed adjustment using slack reclamation in multiprocessor real-time systems". In: *IEEE Transactions on Parallel and Distributed Systems (TPDS)* 14.7 (2003), pp. 686–700. DOI: 10.1109/TPDS.2003.1214320.

91. Amit Kumar Singh, Anup Das, and Akash Kumar. "Energy Optimization by Exploiting Execution Slacks in Streaming Applications on Multiprocessor Systems". In: *Proc. on Design Automation Conference (DAC)*. 2013.
92. Longxin Zhang et al. "Joint optimization of energy efficiency and system reliability for precedence constrained tasks in heterogeneous systems". In: *International Journal of Electrical Power & Energy Systems* 78 (2016), pp. 499–512.
93. A. Martins et al. "Runtime energy management under real-time constraints in MPSoCs". In: *Proc. of IEEE International Symposium on Circuits and Systems (ISCAS)*. 2017, pp. 1–4.
94. Z. Guo et al. "Energy-Efficient Real-Time Scheduling of DAGs on Clustered Multi-Core Platforms". In: *Proc. IEEE Real-Time and Embedded Technology and Applications Symposium (RTAS)*. Apr. 2019, pp. 156–168. DOI: 10.1109/RTAS.2019.00021.
95. V. Chaturvedi et al. "Thermal-aware task scheduling for peak temperature minimization under periodic constraint for 3D-MPSoCs". In: *Proc. IEEE International Symposium on Rapid System Prototyping (RSP)*. Oct. 2014, pp. 107–113. DOI: 10.1109/RSP.2014.6966900.
96. R. Kabir and B. Izadi. "Temperature and energy aware scheduling of heterogeneous processors". In: *Proc. International Conference on Contemporary Computing (IC3)*. Aug. 2016, pp. 1–7. DOI: 10.1109/IC3.2016.7880199.
97. M. Bao et al. "On-line thermal aware dynamic voltage scaling for energy optimization with frequency/temperature dependency consideration". In: *Proc. ACM/IEEE Design Automation Conference (DAC)*. July 2009, pp. 490–495.
98. Hyejeong Hong et al. "Thermal-aware dynamic voltage frequency scaling for many-core processors under process variations". In: *IEICE Electronics Express* 10.14 (2013).
99. Thidapat Chantem, X Sharon Hu, and Robert P Dick. "Temperature-aware scheduling and assignment for hard real-time applications on MPSoCs". In: *IEEE Transactions on Very Large Scale Integration (VLSI) Systems* 19.10 (2011), pp. 1884–1897.
100. M. Qiu et al. "Peak temperature minimization for embedded systems with DVS transition overhead consideration". In: *Proc. International Conference on High Performance Computing and Communication (HPCC) & International Conference on Embedded Software and System ICESS*. June 2012, pp. 477–484.

Chapter 3
Bounding Time in Mixed-Criticality Systems

As a result of MC application analysis, determining the optimally low WCET for each HC task is a challenge in MC system design due to its role in improving the timing behavior of these systems and QoS (scheduling greater number of LC tasks). Determining the high values for low WCETs (i.e., the gap between the low and high WCETs is small) can minimize the number of mode switches but underutilize the processor (i.e., reduce the processor utilization, at run-time) due to the scheduling of fewer tasks. Indeed, this is overly pessimistic because, as shown in Fig. 1.1, tasks would be executed with less likelihood up to observed or actual WCET. On the other hand, the utilization can be maximized, and more tasks, especially LC tasks, are guaranteed to be scheduled in a processor at design-time by determining the low values for low WCETs (i.e., the gap is large), but with a high number of mode switches, and consequently, drop more LC tasks at run-time due to inefficient, low WCET determination for HC tasks, which is not desirable. Most state-of-the-art research works such as [1–3] have presented static approaches to determine low WCETs, which are set as a percentage of the high WCETs. However, as shown in Fig. 1.1, most tasks' execution time is close to Average-Case Execution Time (ACET). Furthermore, most studies have not analyzed the probability of exceeding the low WCETs in system design. To this end, this chapter first aims to adjust the low WCET in Sect. 3.1, which can provide a reasonable trade-off between the number of scheduled LC tasks at design-time and the probability of mode switching at run-time to improve the system utilization and QoS.

However, setting the constant low WCETs for tasks in LO mode, which remain unchanged during run-time, can cause significant performance loss for LC tasks or processor underutilization if the low WCETs are not close to AET. Therefore, we propose *ADAPTIVE* in Sect. 3.2 to determine the low WCETs for MC tasks at run-time based on the behavioral system changes while making a trade-off between the QoS, utilization, and mode switches.

The remainder of the chapter is organized as follows. At first, we propose *BOT-MICS*, in Sect. 3.1, in which a motivational example is presented in Sect. 3.1.1.

B. Ranjbar et al., *Quality-of-Service Aware Design and Management of Embedded Mixed-Criticality Systems*, https://doi.org/10.1007/978-3-031-38960-3_3

Then, we present our scheme for determining the low WCETs, estimating ACET to be used in determining the WCETs, and the scheduling policy and optimization problem in Sect. 3.1.2. At the end of this section, we analyze the experiments in Sect. 3.1.3. Then, the run-time approach, *ADAPTIVE*, is proposed in Sect. 3.2, in which the motivational example and the detail of the proposed scheme are explained in Sects. 3.2.1 and 3.2.2, respectively. The experiments with real-life benchmarks and synthetic task sets are finally evaluated in Sect. 3.2.3.

3.1 BOT-MICS: A Design-Time WCET Adjustment Approach

This section first proposes a novel scheme based on the *Chebyshev theorem* [4] for MC systems to determine the appropriate low WCETs for tasks. The *Chebyshev theorem* provides a general bound for all tasks with any distribution, which is pessimistic. To this end, we then propose a second approach to determine tighter execution time bounds for HC tasks. In this approach, we analyze the execution time distribution of each task and fit a known distribution curve to it. Then we use the Cumulative Distribution Function (CDF) of the known distribution to provide a tight bound for the probability of task overrunning and, consequently, determine the low WCET for that task. Then, the schedulability test and optimization problem based on the newly proposed schemes are discussed.

The main contributions of *BOT-MICS* can be listed as follows:

- Introducing a novel scheme to obtain the low WCETs by the *Chebyshev theorem* in MC systems and showing the relation between the low WCETs and mode switching probability
- Determining the number of adequate samples, for computing ACET and standard deviation
- Representing the tighter execution time bound and more realistic overrunning probability based on the applications' distribution time feature
- Formulating and solving an optimization problem for improving the resource utilization and reducing the mode switching probability using GA
- Evaluating our proposed scheme for various state-of-the-art MC systems to investigate their timing behavior with real benchmarks on a real board, ODROID XU4

3.1.1 Motivational Example

In order to motivate what we have stated, we present an example, in which we executed 20,000 instances of five real-world applications, and their ACETs and WCETs in terms of CPU clock cycle are presented in Table 3.1. $WCET^{HI}$ of each application is determined by OTAWA [5]. For each application, Table 3.1 also shows how many instances violate their $WCET^{LO}$ when it is set to ACET, or fraction ($\frac{1}{4}, \frac{1}{8}, \frac{1}{16}, \frac{1}{32}, \frac{1}{64}$) of the $WCET^{HI}$ [6–8]. The important point that the table shows is by increasing the size of inputs to an application, the ACET and $WCET^{HI}$ growth are not the same. For instance, the growth of $WCET^{HI}$ and ACET for ‹qsort›, a known algorithm for sorting arrays, is O(k^2) and O($k \log k$), respectively, where k is the size of the input array. Therefore, the $WCET^{HI}$ of ‹qsort› application for three different array sizes with 10, 100, and 10,000 elements are 8.1, 22.7, and 59.0 times higher than the ACET of them, respectively. This table shows that $WCET^{HI}$ is not an appropriate parameter to set $WCET^{LO}$. For example, by setting $WCET^{LO}$ to $\frac{WCET^{HI}}{16}$, the mode switching probability for ‹edge›, and ‹qsort-10› is more than 99%, while for ‹smooth›, ‹epic›, ‹qsort-100›, and ‹qsort-10000›, it is less than 2%. On the other hand, when the $WCET^{LO}$ is equal to ACET, the mode switching probability is between 43 and 55% for all applications. So, based on the results in Table 3.1, we can conclude that the mode switching probability is more consistent when the $WCET^{LO}$ is estimated based on ACET, rather than $WCET^{HI}$. However, simply setting $WCET^{LO}$ equal to ACET leads to many system mode changes (almost half of the instances).

To this end, we introduce a scheme that provides a general formula to choose a suitable $WCET^{LO}$ based on ACET to improve the utilization of the system. This approach makes a reasonable trade-off between the mode switching probability and the time that a core becomes idle because of the gap between its AET and the $WCET^{LO}$.

3.1.2 BOT-MICS in Detail

3.1.2.1 Determining Low WCET and Overrunning Probability

Determining the appropriate $WCET^{LO}$ for HC tasks is a major design challenge for MC systems. The proposed scheme designs the MC systems and analyzes the MC tasks of the application in the offline phase. Based on the analysis results, the scheme chooses a suitable $WCET^{LO}$ for each HC task based on their ACET, which improves the number of scheduled LC tasks due to the big gap between the ACET and WCET. To determine $WCET^{LO}$, we introduce the following theorem based on the *Chebyshev theorem*. Note that, the *Chebyshev theorem* is a technique for bounding a tail distribution, which is used for estimating the failure probability and also establishing high probability bounds. In fact, it determines where most of the data samples fall within a distribution. Note that this theorem disregards how the data are distributed. By knowing only the mean (ACET in this book) and standard deviation of data samples, this theorem claims that a certain fraction of these data is less than a certain distance from the mean [4]. In the following, we discuss how this claim helps and how this theorem is employed to estimate $WCET^{LO}$ in MC systems.

Table 3.1 Comparison between ACET and WCET of different applications

App	ACET (cycle)	$WCET^{HI}$ (cycle)	Standard deviation (cycle)	% of samples that overruns if the $WCET^{LO}$ is set to					
				$ACET$	$\frac{WCET^{HI}}{4}$	$\frac{WCET^{HI}}{8}$	$\frac{WCET^{HI}}{16}$	$\frac{WCET^{HI}}{32}$	$\frac{WCET^{HI}}{64}$
qsort-10	2.3×10^2	1.9×10^3	3.9×10^1	50.52	0.00	45.52	99.98	100.00	100.00
qsort-100	1.8×10^4	4.1×10^5	1.2×10^3	50.22	0.00	0.02	0.02	99.98	99.98
qsort-10000	1.8×10^8	1.0×10^{10}	1.1×10^6	43.86	0.00	0.00	0.02	0.02	99.98
Corner	5.6×10^5	9.4×10^6	6.2×10^4	53.27	0.00	0.00	47.71	100.00	100.00
Edge	9.8×10^5	1.1×10^7	1.1×10^5	54.88	0.00	0.00	99.84	100.00	100.00
Smooth	1.9×10^7	4.9×10^8	5.1×10^6	54.31	0.00	0.00	1.41	78.85	97.25
Epic	1.1×10^7	7.0×10^8	1.9×10^6	52.85	0.00	0.00	0.00	0.00	52.20

Theorem 1 *Given a task τ_i, for any positive integer n, the rate at which the execution time exceeds the value $(ACET_i + n \times \sigma_i)$ for task τ_i is bounded with $\frac{1}{1+n^2}$.*

Hence, by considering the Chebyshev theorem (presented below in detail), n can be any positive integer value. However, in our proposed method, it plays an important role to draw a trade-off between determining the $WCET^{LO}$ values and the probability of mode switching. We explain its role after formulating these two parameters.

Proof We use the *Chebyshev theorem* to prove *Theorem 1*:
***One-Sided Chebyshev** [4]* For any nonnegative random variable X, if $E[X]$ is the mean and $Var = \sigma^2$ is its variance, then, for any positive real number $a > 0$, we have the theorem (3.1):

$$Pr[X - E[X] \geq a] \leq \frac{\sigma^2}{\sigma^2 + a^2} \tag{3.1}$$

In this theorem, if a is equivalent to $n \times \sigma$ $(a \equiv n \times \sigma)$:

$$Pr[X - E[X] \geq n \times \sigma] \leq \frac{1}{1 + n^2} \tag{3.2}$$

Now, assuming m samples of task τ_i $(j_{i,1}, j_{i,2}, \ldots, j_{i,m})$ with execution time $C_{i,1}, C_{i,2}, \ldots, C_{i,m}$, the expected value $E[X]$ of task τ_i is:

$$E[X] = ACET_i = \frac{1}{m} \sum_{j=1}^{j=m} C_{i,j} \tag{3.3}$$

Here, $j_{i,k}$ represents job k of task τ_i, and each job $J_{i,k}$ has the execution time value of $C_{i,k}$.

By using the expected value $ACET_i$ (we present how to compute $ACET$ for each task τ_i in the next subsection), the standard deviation of execution time, σ_i, for task τ_i is calculated as follows:

$$\sigma_i = \sqrt{\frac{1}{m} \sum_{j=1}^{j=m} (C_{i,j} - ACET_i)^2} \tag{3.4}$$

If the execution time of a task is considered as a random variable, by using the *Chebyshev theorem*, we can show that less than $\frac{1}{1+n^2}$ of samples have higher execution time than n standard deviation $(n \times \sigma)$ of the mean execution time (*ACET*=E[X]):

$$Pr[X \geq ACET_i + n \times \sigma_i] \leq \frac{1}{1 + n^2} \tag{3.5}$$

Therefore, the rate of exceeding the execution time level $(ACET_i + n \times \sigma_i)$ for task τ_i is bounded with $\frac{1}{1+n^2}$. ■

This theorem provides a general upper bound on the probability of exceeding any arbitrary execution time level for any task, independent of its distribution. To determine $WCET^{LO}$, the *Chebyshev theorem* can be applied, which requires mean ($ACET$, that we discuss later how to compute it in the next subsection) and standard deviation of the execution time (σ) of each task:

$$C_i^{LO} = WCET_i^{LO} = ACET_i + n_i \times \sigma_i \tag{3.6}$$

Parameter n should be set very carefully because a large value of n reduces the number of scheduled tasks in LO mode, and a small n increases the probability of mode switching $P_i^{MS} = \frac{1}{1+n^2}$. In Sect. 3.1.3, we evaluate the impact of different values of n in computing the $WCET^{LO}$ and the probability of system mode switching (P_{sys}^{MS}).

In addition, since the value of $WCET^{LO}$ is based on the ACET, we need to calculate the average execution time for each task. In general, it is hard to achieve the real mean (μ) with all possible samples of tasks [9], so we discuss a method to estimate the empirical mean ($\hat{\mu}$) with the minimum number of samples.

3.1.2.2 ACET Estimation and Its Minimum Required Samples

In order to calculate the $WCET_i^{LO}$ for each task τ_i, we explain how to estimate the $ACET_i$. Therefore, we need to determine how many samples (m) are required for ACET estimation. We present the estimation by the probability $1 - \delta$ and ϵ error as follows.

Theorem 2 *For any task τ_i, consider m as the number of samples; if $m \geq ln(\frac{2}{\delta})\frac{(WCET_i^{HI})^2}{2(\epsilon \times \mu)^2}$, there is an (ϵ, δ)-approximation for computing $ACET$ of task τ_i.*

Proof Considering m samples of task τ_i, where $C_{i,1}, C_{i,2}, \ldots, C_{i,m}$ are their execution times. Then, the empirical mean ($\hat{\mu}$) for task τ_i is computed as Eq. (3.7):

$$\hat{\mu} = \frac{1}{m}\sum_{j=1}^{j=m} C_{i,j} \tag{3.7}$$

To prove *Theorem 2*, we use the *Hoeffding bound* theorem [10] to approximate the real mean (μ). Note that the *Hoeffding bound* theorem provides an upper bound on the probability that the sum of random variables with a bounded range deviates from its expected value by more than a certain value [10, 11]. The execution time of a sample is an independent random variable because the execution time of one sample does not affect other samples' execution time.

Hoeffding Bound Let $C_{i,1}, C_{i,2}, \ldots, C_{i,m}$ be independent random variables which are bounded by an interval $[a, b]$, and then:

$$Pr[|\hat{\mu} - E[\hat{\mu}]| \geq \epsilon] \leq 2e^{\left(-\frac{2m\epsilon^2}{(b-a)^2}\right)} \tag{3.8}$$

The *Hoeffding* theorem bounds $Pr[|\hat{\mu} - \mu| \geq \epsilon]$ by using the fact that $E[\hat{\mu}] = \sum_{j=1}^{j=m} E[C_{i,j}] = \mu$. Thus, it can estimate the real mean μ with ϵ error. Based on the *Hoeffding* theorem, the execution time of each sample must be bounded by an interval $[a, b]$. The upper bound execution time of the task's samples is the high WCET of that task ($WCET^{HI}$), so the execution time of each instance is bounded by $[0, WCET_i^{HI}]$ interval. Therefore, $b - a \leq WCET_i^{HI}$. If we consider $\epsilon = \epsilon^* \times \mu$, Eq. (3.8) is written as:

$$Pr[|\hat{\mu} - \mu| \geq \epsilon^* \times \mu] \leq 2e^{\left(-\frac{2m(\epsilon^* \times \mu)^2}{(WCET_i^{HI})^2}\right)} \tag{3.9}$$

In order to estimate the real mean with the minimum number of samples, we use a definition of (ϵ, δ)-*Approximation* [10].

(ϵ, δ)-***Approximation*** An algorithm gives an (ϵ, δ)-approximation for the input value V if the output X of this algorithm satisfies the following inequality. In fact, output X approximates input V with probability $1 - \delta$ and ϵ error:

$$Pr[[X - V] \leq \epsilon V] \geq 1 - \delta \Leftrightarrow Pr[[X - V] \geq \epsilon V] \leq \delta \tag{3.10}$$

By using this definition and Eq. (3.9), we present the following equation to achieve a $(1 - \delta)$ confidence for the correctness of such an approximation:

$$2e^{\left(-\frac{2m(\epsilon \times \mu)^2}{(WCET_i^{HI})^2}\right)} \leq \delta \Longrightarrow ln\left(\frac{2}{\delta}\right)\frac{(WCET_i^{HI})^2}{2(\epsilon \times \mu)^2} \leq m \tag{3.11}$$

This equation shows that with $m \geq ln(\frac{2}{\delta})\frac{(WCET_i^{HI})^2}{2(\epsilon \times \mu)^2}$ instances, $\hat{\mu}$ is an (ϵ, δ)-approximation for μ:

$$Pr[|\hat{\mu} - \mu| \geq \epsilon \times \mu] \leq \delta \tag{3.12}$$

■

3.1.2.3 Determining a Tight Execution Time Bound

Equation (3.5) presents a general theorem that is applied to any time distribution of tasks. Therefore, it might not provide a tight upper bound for the probability of mode switching. For example, if we consider $n = 0$ ($WCET_i^{LO} = ACET_i$),

Table 3.2 The effect of varying n on the overrunning of different tasks from MiBench suite [12], under the proposed Chebyshev-based scheme and experiments

	Chebyshev	Bitcount	qsort	Matrix-mult	Smooth	Corner
$n = 0$	100.00%	43.31%	33.92%	42.33%	33.47%	7.96%
$n = 1$	50.00%	8.87%	6.30%	16.26%	19.95%	4.95%
$n = 2$	20.00%	3.68%	4.37%	4.18%	4.92%	3.98%
$n = 3$	10.00%	0.92%	2.33%	0.91%	1.43%	3.08%
$n = 4$	5.88%	0.71%	1.12%	0.22%	0.39%	2.22%

the rate of exceeding $ACET_i$ for task τ_i is bounded with 100% by the *Chebyshev theorem*. It means the execution time of all samples of task τ_i might be more than $ACET_i$, which is not true for most distributions. Although it is not wrong, it does not provide a piece of useful information. Table 3.2 shows the percentage of overruns for five different applications, from MiBench suite [12] through experiments and our analysis, *Chebyshev*-based scheme. As shown, the proposed scheme can provide an upper bound which is valid for any execution time distribution. However, this scheme gives a high and loose upper bound for many applications. As an example, the percentage of overruns in experiments for ‹corner› application is 7.96% when $n = 0$, while according to our scheme, it is estimated to be 100%.

Since in our case, the tasks' execution time distribution for some applications is known, we propose another scheme, an alternative one, to determine the tighter execution time bounds. As we discuss further, the determined WCETs would be more realistic, which cause the method to be more scalable. Note that this method might help for better scale to multiple criticality levels and thus, better management of mode switches. To preset the tighter execution bounds, we execute several benchmarks on a real board (we discuss the details in Sect. 3.1.3) and investigate their time distributions. Figure 3.1a depicts the execution time distribution of four applications, from MiBench suite [12]. The distribution curve of these applications is very similar to existing known probability distributions. Therefore, we fitted these applications with well-known distributions. We used features of those distributions like Probability Density Function (PDF) and CDF to estimate a tighter upper bound for mode switching. A distribution's CDF shows the probability that the execution time of an instance is less than or equal to a certain value, which we can consider as the low WCET. Figure 3.1b shows the fitted PDF and Fig. 3.1c shows the empirical and fitted CDF for four benchmarks. Since the probability of each task overrunning is important in our proposed method, we use the CDF formula (based on the best-fitted distribution) as $(1 - P_i^{MS})$ in our proposed method to find a tighter bound.

To identify the best-fitted distribution for the applications' execution time data, we have considered 16 different data distributions such as *Normal*, *Burr*, *Gamma*, *t*, *Weibull*, *Lognormal*, etc. We evaluate the distributions' efficacy using Kolmogorov-Smirnov's (K-S) fitness metric [13], which is a commonly used technique. We select the top three distributions to implement the corresponding fitness functions for each application. As an example, Fig. 3.2a shows the density of the top three distributions for the ‹insertion-sort› application, which are Burr, t, and Weibull distributions.

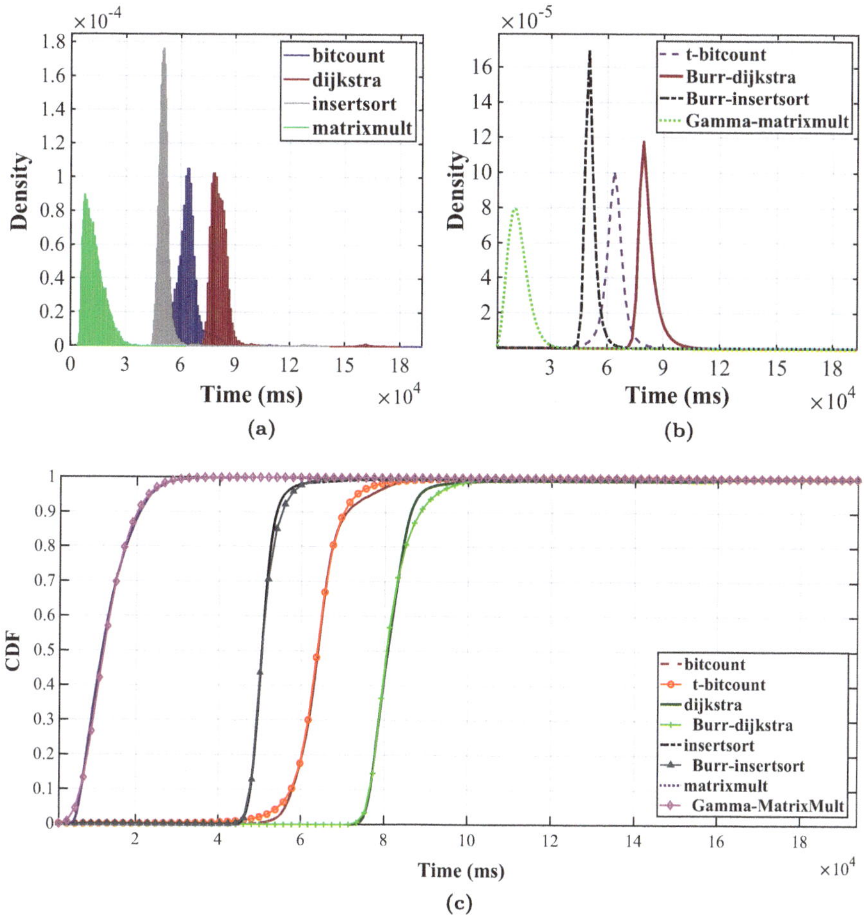

Fig. 3.1 Empirical execution time distributions and fitted distributions. (**a**) Empirical time distribution. (**b**) Fitted distribution. (**c**) Empirical and fitted CDF

Besides, to see how well a distribution fits data, we show how empirical data is distributed compared with a fitted distribution. Therefore, by using probability-probability (p-p) plot [14], we show two CDFs against each other. Figure 3.2b, c, and d show it for the empirical and fitted data for the top three distributions. As shown, the Burr distribution (Fig. 3.2b) is more matched between the observed and theoretical cumulative distributions compared to t distributions (Fig. 3.2c) and Weibull (Fig. 3.2d) distribution.

In the end, to compute a tighter probability of task overrunning ($Prob_i^M$) based on n, $ACET$, and σ, instead of using Eq. (3.5), the CDF of the determined distribution ($F_i(t)$) is used as $Prob_i^{MS} = 1 - F_i(ACET_i + n \times \sigma_i)$.

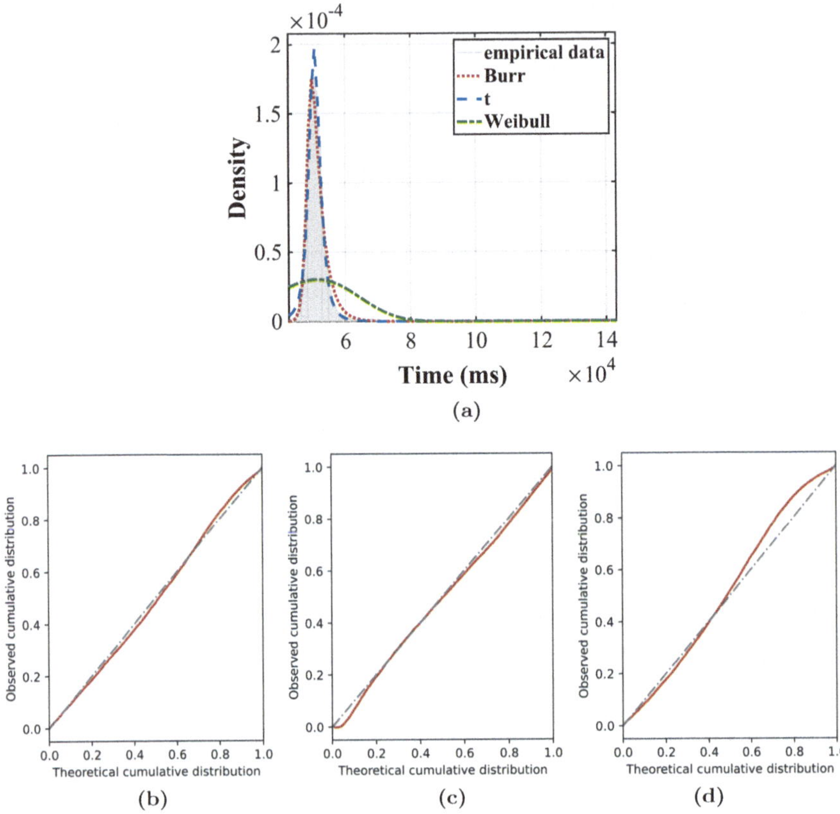

Fig. 3.2 PDF and CDF of top three distributions for insertsort benchmark. (**a**) Top three distributions' PDF for insertsort. (**b**) CDF of Bur distribution. (**c**) CDF of t distribution. (**d**) CDF of Weibull distribution

3.1.2.4 Task Schedulability Analysis

In this subsection, we analyze the task schedulability and present the conditions based on the new formula of $WCET^{LO}$, determined in previous subsections. To schedule MC tasks in the uni-processor, we apply the existing MC scheduling technique, EDF-VD algorithm, which has been used in many studies since the last decade [1, 6, 7]. Here, when the system switches to the HI mode, all LC tasks are dropped. If U_l^k denotes total utilization of tasks with the same criticality level l in the mode k, then:

$$U_{HC}^{LO} = \sum_{\zeta_i = HC} \frac{ACET_i + n_i \times \sigma_i}{T_i} \quad \text{and} \quad U_{HC}^{HI} = \sum_{\zeta_i = HC} \frac{WCET_i^{HI}}{T_i} \tag{3.13}$$

A suitable $WCET^{LO}$ for each HC task τ_i can be achieved by choosing the optimum n_i (used in Eq. (3.6)). The optimum n_i must be determined to minimize the mode switching probability and maximize resource utilization. To solve this, we formulate the optimization problem to find the optimum n_i for each task τ_i and determine its $WCET_i^{LO}$. Furthermore, Eq. (3.14) must be satisfied to guarantee schedulability by EDF-VD at run-time [1]. Equation (3.14) presents the necessary and sufficient conditions to guarantee the task schedulability in both LO mode and HI mode and meeting deadlines of running tasks even if the system switches to the HI mode [1]:

$$U_{HC}^{LO} + U_{LC}^{LO} \leq 1 \quad \text{and} \quad U_{HC}^{HI} + \frac{U_{HC}^{LO} \times U_{LC}^{LO}}{1 - U_{LC}^{LO}} \leq 1 \tag{3.14}$$

3.1.2.5 Optimization Problem Formulation

In order to formulate the optimization problem based on the two objectives (mode switching probability and system utilization), we first identify the variables and constraints for better understanding. For each task τ_i, $ACET_i$, $WCET_i^{HI}$, T_i (period), and σ_i (standard variation) are constant. $WCET_i^{LO}$ is variable, which is computed based on the variable n_i (introduced in the beginning of the subsection). These constant parameters and variables are used to compute the objectives, mode switching probability, and system utilization. In order to optimize these objectives and find the optimum value for n_i, we first present the constraints and then formulate the objectives as follows.

Execution Time Constraint $WCET_i^{LO}$ of each HC task τ_i must not be more than $WCET_i^{HI}$:

$$ACET_i + n_i \times \sigma_i \leq WCET_i^{HI} \tag{3.15}$$

There are two main objectives to optimize the system:

Objective 1 (Mode Switching Probability) If the LC tasks are dropped frequently due to the HC tasks' overrunning, it may negatively impact the performance or functionality of MC systems. Therefore, one of the most significant objectives is the minimization of mode switching probability. Let P_{Sys}^{MS} denote the probability of system mode switching. If P_{Sys}^{noMS} is the probability that no HC task overruns and consequently, no mode switch happens, then $P_{Sys}^{MS} = 1 - P_{Sys}^{noMS}$. Since tasks are independent, P_{Sys}^{MS} is computed as shown in Eq. (3.16), where P_i^{MS} is the probability of task overrunning for task τ_i. According to our discussion, $P_i^{MS} = \frac{1}{1+n_i^2}$. The higher the n_i, the less the mode switching probability:

$$P_{Sys}^{MS} = 1 - \prod_{\zeta_i \in HC} \left(1 - P_i^{MS}\right) = 1 - \prod_{\zeta_i \in HC} \left(1 - \frac{1}{1 + n_i^2}\right) \tag{3.16}$$

Objective 2 (Resource Utilization) The second objective is to improve the resource utilization by a significant gain in terms of the utilization that can be allocated to LC tasks in the LO mode (U_{LC}^{LO}). Although maximizing U_{LC}^{LO} is desired, it is upper-bounded by the schedulability constraints, which can be derived from Eq. (3.14). Equation (3.17) presents the condition to guarantee the task schedulability in the LO mode under the EDF-VD algorithm. In addition, as mentioned in the previous subsection, Eq. (3.18) shows the condition for guaranteeing the task schedulability in the HI mode and mode switching [15, 16]. In this equation, the maximum amount of U_{LC}^{LO} depends on the values of n_i for each HC task. The lower the n_i, the higher the U_{LC}^{LO}. Therefore, the second objective can be bounded as follows:

$$U_{LC}^{LO} \le 1 - U_{HC}^{LO} = 1 - \sum_{\zeta_i = HC} \frac{ACET_i + n_i \times \sigma_i}{T_i} \tag{3.17}$$

$$U_{LC}^{LO} \le \frac{1 - U_{HC}^{HI}}{1 - U_{HC}^{HI} + U_{HC}^{LO}} = \frac{1 - U_{HC}^{HI}}{1 - U_{HC}^{HI} + \sum_{\zeta_i = HC} \frac{ACET_i + n_i \times \sigma_i}{T_i}} \tag{3.18}$$

Hence, if P_{Sys}^{MS}=1, it means the system is always in the HI mode, and all LC tasks are always dropped. If P_{Sys}^{MS}=0, it implies all LC tasks are always executed with no dropping. Therefore, by having these two objectives, we maximize the following equation:

$$maximize\left\{\left(1 - P_{Sys}^{MS}\right) \times U_{LC}^{LO}\right\} \tag{3.19}$$

Problem Solving: Derivation-Based Optimization In order to optimize the two objectives of mode switching probability and utilization, the optimum value of n_i must be obtained for each task τ_i. If the uniform n is considered for all tasks to compute the $WCET_i^{LO}$, we can obtain the optimum n by finding the derivation of both objectives. Using the method of the second derivative helps to find the largest or smallest value of a function, where the derivative equals zero. Further details, on how the derivative works to find the optimum value, are provided in the result section (Sect. 3.1.3.2) by an example. However, obtaining the uniform optimum n for all tasks is not fair and tasks have different time distributions. Table 3.3 shows the minimum value of n for some benchmarks of MiBench suite, where $WCET_i^{LO} = ACET_i + n \times \sigma \ge WCET_i^{HI}$. Due to having different time distributions of tasks, choosing the uniform n causes the system's objectives to not optimize well and precisely. As a result, optimization techniques that can handle

Table 3.3 The minimum value of n in $WCET_i^{LO} \geq WCET_i^{HI}$ for different tasks

	FFT	qsort	dijkstra	Corner	Edge	Smooth	Epic	Bitcount
n	60	17	12	11	27	8	7	19

nonuniform values of n_i across different tasks and can scale effectively with the increasing number of tasks in the system are necessary.

Problem Solving: GA-Based Optimization Global optimization methods based on randomized algorithms have been used extensively in system-level design space exploration for QoS improvement in embedded systems [17]. In our current work, we use GA for solving the maximization problem shown in Eq. (3.19). GA involves using randomized search methods based on the principles of natural evolution and genetics.

It is important to mention that Mixed Integer Linear Programming (MILP) can be used as an alternative to GA for optimization. However, the problem formulation of MILP is much more complex compared to that of GA, which allows a simpler implementation of the fitness function. Although GA has a lack of optimality guarantees, MILP also does not scale very well with the number of integer variables. So, an increased number of integer (and real) variables resulting from a large number of tasks—n_i and support variables—in an MILP formulation can increase the complexity considerably. Most state-of-the-art tools for solving MILP problems also provide a time-bound *best-effort* solution for complex problems. Further, for the distribution-aware optimization for real-world tasks, we use a lookup table to search for the closest WCET and probability of mode switching values. Implementing such lookup-based optimizations from real-world observations with standard MILP formulation can be considerably more complex than using GA. It must be noted that the focus of the work is on showing the efficacy of the proposed methodology in providing improved trade-offs between mode switching probability and utilization. While we would ideally prefer optimization methods with guaranteed optimality, the choice of GA was based on the ease of implementation and the support for integrating varying estimation methods–both mathematical and lookup-based. However, MILP formulation for the current research problem can be a suitable topic for further exploration. The encoding approach and GA methods used in our current work include the following:

- *Individual*: An ordered sequence of integer values forms the individual in the population. Each integer in the sequence corresponds to the value of n_i for a task τ_i.
- *Population*: During the optimization, we generate two types of individuals for initializing the population of the first generation of candidate solutions. Firstly, we generate individuals comprising of randomly sampled n_i values from the range $[1, 50]$ for each task τ_i in the benchmark. Secondly, we generate uniform-valued individuals from the same range to ensure that the optimization included uniform values of n_i for each task.

- *Crossover* and *Mutation*: We used two-point crossover for exchanging n_i values among two candidate solutions. During crossover, the configurations of the two randomly selected possible solutions are interchanged. This process forms one of the algorithms that generates new possible solutions (individuals) for the next generation of solutions. In our current problem, this entails interchanging the n_i values of two candidate solutions, selected from the current generation, for a subset of the tasks. Similarly, we used single-point mutation to set the value of n_i for a randomly selected task in the candidate solution to a randomly selected value in the range $[1, 50]$.
- *Selection*: We use tournament selection for choosing the candidate solutions for the population of the next generation. It involves randomly choosing a fixed number of individuals from the current population and selecting the one with the maximum value of $(1 - P_{Sys}^{MS}) \times U_{LC}^{LO}$ for the next generation.
- *Fitness* and *Feasibility*: Eqs. (3.16)–(3.18) were used to evaluate the fitness ($(1 - P_{Sys}^{MS}) \times U_{LC}^{LO}$) of each candidate solution. Similarly, Eqs. (3.14) and (3.15) were used to determine the feasibility of each candidate solution.

3.1.3 Evaluation

Now, we present the experiments to evaluate the effectiveness of our proposed scheme in terms of utilization, schedulability, and mode switching probability.

3.1.3.1 Evaluation with Real-Life Benchmarks at Run-Time

Evaluation Setup To evaluate our scheme, we conducted some experiments on the ODROID XU4 board powered by ARM, which has the big.LITTLE architecture, with four Cortex A15 (big) and four Cortex A7 (LITTLE) cores. We use the LITTLE cores with the maximum frequency of 1.4 GHz, for doing the experiments.

To evaluate our scheme by real benchmarks, we use various benchmarks from MiBench benchmark suite [12] such as automotive, network, and telecomm. and from AXBench [18] such as matrix-multiplier. We execute each benchmark with different inputs on the ODROID-XU4 board, to achieve their execution times. Table 3.4 shows the high WCET, ACET, and σ (standard deviation) of these benchmarks.

Minimum Required Samples to Estimate ACETs Figure 3.3 shows the minimum required samples of each benchmark based on *Theorem 2* to estimate ACET, by varying the parameters of *(ϵ, δ)-Approximation*. In fact, as an example in Fig. 3.3a, with 90% confidence, the estimation error of ACET for each benchmark is less than $(\epsilon \times \mu)$, where μ is the real mean. Besides, by decreasing the confidence, the minimum required samples for each benchmark is decreased. It means with more

Table 3.4 Execution time distribution of various benchmarks

Exe. Par. (ms)	Insertsort-10000	Matrix-multiply	qsort-10000	Corner	Edge	Smooth	Epic	Bitcount	dijkstra	FFT
$WCET^{HI}$	753.23	387.67	759.32	51.63	131.47	301.09	230.81	1142.17	1039.98	686.52
$ACET$	51.33	13.05	39.65	0.55	0.94	9.317	2.69	64.73	81.95	6.15
σ	6.38	5.6	5.46	0.71	0.87	5.63	1.98	6.94	8.65	2.12

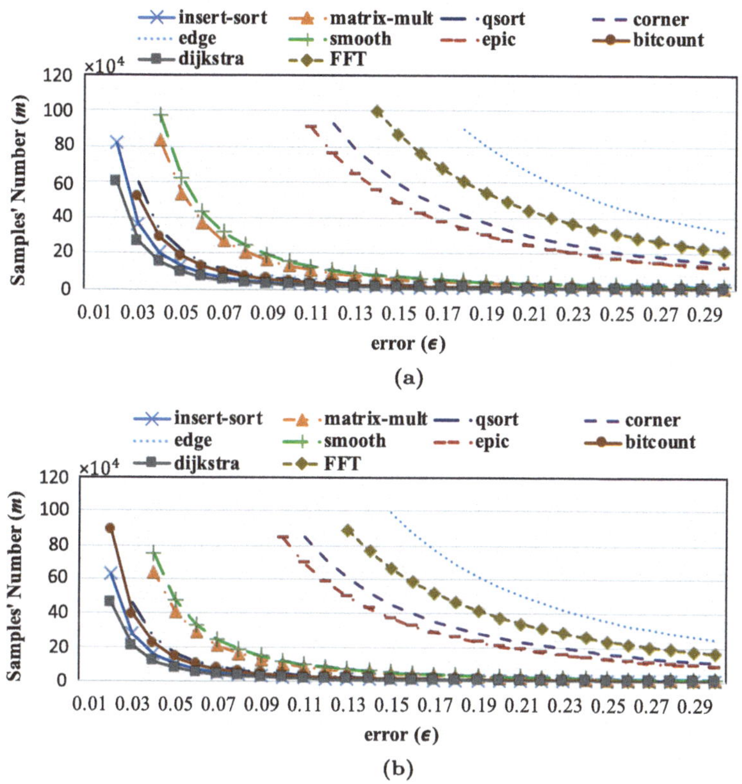

Fig. 3.3 Required number of samples for different benchmarks by varying the error (ϵ) and confidence $(1-\delta)$. (**a**) $1-\delta=0.9$. (**b**) $1-\delta=0.8$

samples, we can say with more confidence that the difference between the estimated average and the real average is less than $\epsilon \times ACET^{real}$.

Investigating MC Systems' Timing Behavior In order to evaluate the proposed approach, we run these benchmarks on a single core. We consider ‹insert-sort›, ‹matrix-mult›, ‹qsort›, ‹bitcount›, ‹dijkstra›, and ‹FFT› as HC tasks and ‹corner›,

Table 3.5 System performance in both design-time and run-time phases, for different scenarios

	Dropped LC jobs (%)	$max(U_{LC}^{LO})$	P_{Sys}^{MS}	$max(U_{LC}^{LO}) \times (1 - P_{Sys}^{MS})$
[1] $\lambda = \frac{1}{2}$	0	44.7%	0.24%	0.446
[1] $\lambda = \frac{1}{4}$	0	61.78%	1.21%	0.610
[1] $\lambda = \frac{1}{8}$	0.33%	76.37%	10.23%	0.686
[1] $\lambda = \frac{1}{16}$	39.29%	86.61%	92.02%	0.069
Chebyshev	0.06%	77.01%	7.1%	0.715
Dist. analyt.	0	84.31%	2.25%	0.824

‹edge›, ‹smooth›, and ‹epic› as LC tasks. We compute the low WCET for each HC task based on the three policies—our scheme under the *Chebyshev theorem*, our scheme under distribution analysis, and the fraction analysis. In order to specify what the fraction analysis is, most of the state-of-the-art approaches have defined a fraction of $WCET^{HI}$ as $WCET^{LO}$. For example, if we define $\lambda = \frac{WCET^{LO}}{WCET^{HI}}$, researchers in [2] have considered $\lambda \in [\frac{1}{2.5}, \frac{1}{1.5}]$ in their experiments. In [1], two different ranges for λ have been considered, $\lambda \in [\frac{1}{4}, 1]$ and $\lambda \in [\frac{1}{8}, 1]$. Researchers in [6] have considered the amount of $\lambda = \{\frac{1}{16}, \frac{1}{8}, \frac{1}{4}, \frac{1}{2}, 1\}$. Since all papers have the same policy to determine $WCET^{LO}$, we choose [1] as a representative of these approaches.

For these real tasks, including both LC and HC tasks, the system with $\lambda = 1$ has the utilization of more than one in the worst-case scenario, and then it is not schedulable. Therefore, we only consider the amount of λ as $\{\frac{1}{16}, \frac{1}{8}, \frac{1}{4}, \frac{1}{2}\}$. Here, we investigate the system at run-time for 1000 hyper-period of tasks and see how often these tasks exceed their $WCET^{LO}$ under various policies and the system has to switch to the HI mode.

By reducing the λ, the low WCET ($WCET_i^{LO}$) for HC tasks decreases, and the system executes more LC tasks. But on the other hand, it causes frequent system switches and more LC tasks dropping during run-time, leading to lower QoS. As an example, Table 3.5 shows the maximum total utilization bound that can be assigned to LC tasks at design-time for each scenario and also the percentage of dropped LC tasks due to the system mode switching at run-time. We assume that the system needs to run different instances of each LC tasks (with different input) as much as possible to improve the QoS. As shown in Table 3.5, the *Chebyshev*-based scheme can schedule more LC tasks in the system compared to [1] approach. Although the maximum assigned utilization to LC tasks is almost equal to the scenario of [1] with $\lambda = \frac{1}{8}$, the mode switching probability and the number LC dropped tasks are lower in the *Chebyshev*-based scheme. This is because a fraction of high WCET does not provide any information about how many samples might exceed it. So, setting the low WCET of each task equal to $\lambda = \frac{1}{8}$ of the high WCET of that task is too low for some tasks and too high for others. The optimization goal in the last column of the table shows the fact that the *Chebyshev*-based scheme performs better than the approach of [1] (i.e., the goal metric has a larger value). Besides, the distribution analytics-based scheme improves total utilization by 7.3% compared to the *Chebyshev*-based scheme. It also reduces the mode switching probability by 4.85%. This is because the *Chebyshev* is a general formula that is valid for any distribution, but it is not very optimistic. The value of the optimization goal in the last column of the table also shows this fact. Let us consider the distribution analytics-based scheme with the method of [1] with $\lambda = \frac{1}{16}$ which both have almost the same total utilization. The results show that in the distribution analytics-based, the probability of mode switching and the percentage of dropped LC tasks are 89.75% and 39.29% lower, respectively, which is desirable.

3.1.3.2 Evaluation with Synthetic Task Sets

Task Set Generation and Evaluation Setup In order to further evaluate our scheme, we generated synthetic dual-criticality task sets similar to the state-of-the-art studies [1, 19–21], for various system utilization bounds (U_{bound}) in line with the previous works [1, 8, 19–21], where ($U_{bound} = max(U_{LC}^{LO} + U_{HC}^{LO}, U_{HC}^{HI})$). The algorithm adds tasks to the task set randomly to increase the U_{bound} until it reaches a given threshold. We evaluate different approaches for U_{bound} in the range of [0.05, 1] with steps of 0.05, and for each U_{bound}, 1000 task sets are generated. Here, we consider balanced tasks in terms of criticality levels, i.e., the probability of a generated task being HC is equal to being LC. Besides, inspired by real execution times, presented in Table 3.4, we provide the $WCET^{HI}$, $ACET$ and σ in the range of [52,1142], [0.55,81.95], and [0.71,8.65] ms, respectively, where $WCET^{HI} > ACET$. As a result, the periods of tasks are computed based on the task utilization and $WCET^{HI}$ ($u_i^{HI} = \frac{WCET_i^{HI}}{P_i}$).

The recent advanced features in CAD tools, like MATLAB, Excel, and new libraries in Python, provide several practical ways to find a distribution that fits the best to the data samples. Besides, the probabilistic analysis for distribution fitting is implemented in Python using multiple packages, including scikit-learn. For solving the formulated problem with GA, we set the mutation probability to 0.2 and the crossover probability to 0.8. We also used five individuals in the tournament selection process. The optimization methods were implemented in Python using the DEAP [22] package. In the following, we perform extensive simulations to evaluate the effectiveness of our proposed approach in comparison with the state-of-the-art methods.

Effect of Varying Uniform non Maximum Assigned Utilization to LC Tasks and Mode Switching Probability for a Task Set Example In this section, we evaluate the effects of varying the parameter n, used to determine $WCET^{LO}$ for each HC task, on system properties. In this experiment, for the sake of presentation, we considered only one n (uniform) for all HC tasks. However, in further experiments, due

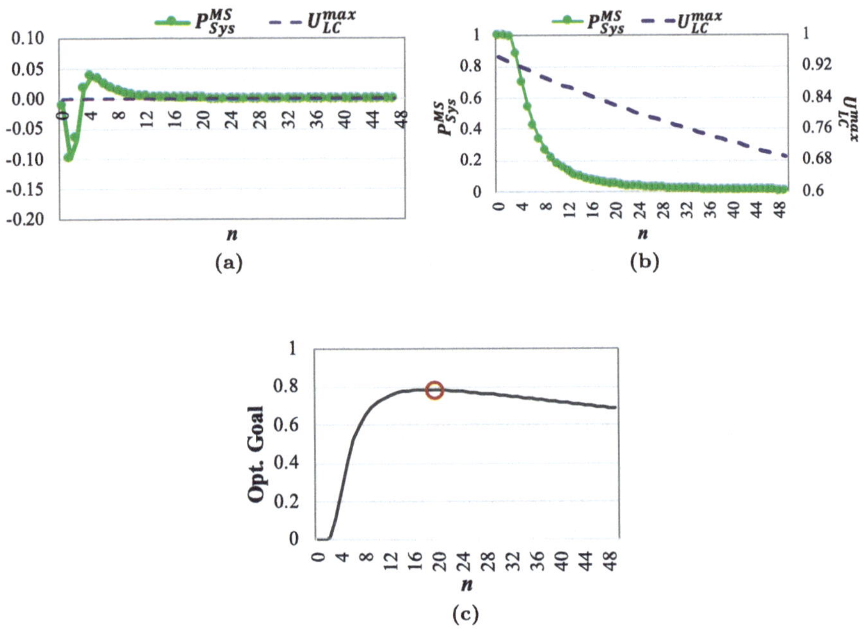

Fig. 3.4 Effect of varying uniform n on maximum assigned utilization to LC tasks and mode switching probability for an example task set. (**a**) Second derivation of system properties. (**b**) P_{sys}^{MS} and $max(U_{LC}^{LO})$. (**c**) Objective function

to our explanation in Section "Problem Solving: Derivation-Based Optimization", we find an independent n for each task with the help of the GA. As mentioned, we improve resource utilization by a significant utilization that can be allocated to LC tasks in the LO mode. Figure 3.4 shows the results for an example task set with $U_{HC}^{HI} = 0.84$. First, we show the results, solved by derivation-based optimization in Fig. 3.4a, and then by GA-based optimization in Fig. 3.4b and c.

Figure 3.4a shows the second derivative of utilization and mode switching probability for the task set. The figure shows that the second derivative of utilization is almost zero all the time, while the second derivative of mode switching probability becomes almost zero for $n \geq 18$. Therefore, we can conclude that the mode switching probability impacts more on obtaining the optimum value of n. To show how effective the derivation-based optimization is, we solve the problem with uniform n by GA-based optimization for this example, shown in Fig. 3.4b and c.

Equation (3.15) shows that by increasing the value of n, the $WCET^{LO}$ of HC tasks and consequently HC tasks' utilization in the LO mode are increased, which reduces the number of scheduled LC tasks at design-time ($max(U_{LC}^{LO})$). On the other hand, Eq. (3.16) shows that by increasing the value of n, the probability of mode switching (P_{sys}^{MS}) is decreased, which means fewer LC tasks are dropped at run-time. Figure 3.4b depicts that, by increasing the value of n, both P_{sys}^{MS} and $max(U_{LC}^{LO})$ are decreased, while to achieve the best utilization, we need to maximize $max(U_{LC}^{LO})$ and minimize P_{sys}^{MS}. Therefore, if the n is set to 5, then P_{sys}^{MS} is equal to 0.54 and $max(U_{LC}^{LO})$ is equal to 0.91. Meanwhile, for $n = 10$, P_{sys}^{MS} is equal to 0.18 and $max(U_{LC}^{LO})$ is equal to 0.88. Indeed, P_{sys}^{MS} is decreased at a great rate by increasing n, compared to $max(U_{LC}^{LO})$ decrements. Now, consider $n = 20$ where $P_{sys}^{MS} = 0.05$ and $max(U_{LC}^{LO}) = 0.82$. It can be seen that the rate of P_{sys}^{MS} reduction is decreased by increasing n, while $max(U_{LC}^{LO})$ reduction rate is very low. Therefore, $max(U_{LC}^{LO})$ becomes more important than P_{sys}^{MS} in this case. We used Eq. (3.19) to find a proper n which makes a trade-off between P_{sys}^{MS} and $max(U_{LC}^{LO})$ and improves the system utilization. Figure 3.4c shows that the optimum n is 18 for our case study task set where $max(U_{LC}^{LO}) = 83\%$ and $P_{sys}^{MS} = 0.06$.

Effect of Varying Uniform n on Maximum Assigned Utilization to LC Tasks and Mode Switching Probability for More Task Sets Now, we evaluate the effects of parameter n and different utilization of HC tasks on system properties in Fig. 3.5, by running 1000 task sets for each utilization point. According to Fig. 3.5a, P_{sys}^{MS} is increased when utilization increases. For example, for a constant $n = 10$, for U_{HC}^{HI} equal to 0.4 and 0.8, P_{sys}^{MS} is 15.47% and 28.43%, respectively. The reason is when utilization of HC tasks is high, more HC tasks are scheduled in the system. Since each HC task has the probability of overrunning, by increasing the number of HC tasks, P_{sys}^{MS} is increased. In addition, we discussed that P_{sys}^{MS} is decreased by increasing n. Figure 3.5b also shows that by increasing U_{HC}^{HI}, there is less opportunity to schedule LC tasks. As a result, the system schedules fewer LC tasks which degrades $max(U_{LC}^{LO})$. As an example, for a constant $n = 10$, if $U_{HC}^{HI} = 0.4$, then $max(U_{LC}^{LO}) = 87.59\%$, and if $U_{HC}^{HI} = 0.8$, then $max(U_{LC}^{LO}) = 53.46\%$. Besides, as mentioned, increasing n causes a decrease in $max(U_{LC}^{LO})$. As a result, by increasing n, P_{sys}^{MS} is reduced (which is desirable), and the LC task utilization and consequently schedulability are also reduced (which is not desirable). Now, if we optimize both P_{sys}^{MS} and assigned utilization to LC tasks, we can find the optimum value of n for HC tasks. Figure 3.5c shows the product of P_{sys}^{MS} and $max(U_{LC}^{LO})$ (Eq. (3.19)), where the optimum n is decreased in general with an increase in U_{HC}^{HI}, to run more tasks in the system.

Comparison with the Other Policies Since applications have different time distributions, choosing the uniform n prevents the system from optimizing its objectives precisely. Therefore, solving the problem with optimization algorithms like GA is the best method to optimize system properties. As a result, in this subsection, we compare the mode switching probability and resource utilization under our proposed scheme with nonuniform n using the GA, and the other policies, used to determine $WCET^{LO}$ and then U_{HC}^{LO}.

Since ACET and σ for each task are known, the system mode switching probability for other policies can be obtained using Eq. (3.6). Figure 3.6 shows the results of comparing different policies and our scheme with the optimum n_i for each task τ_i of

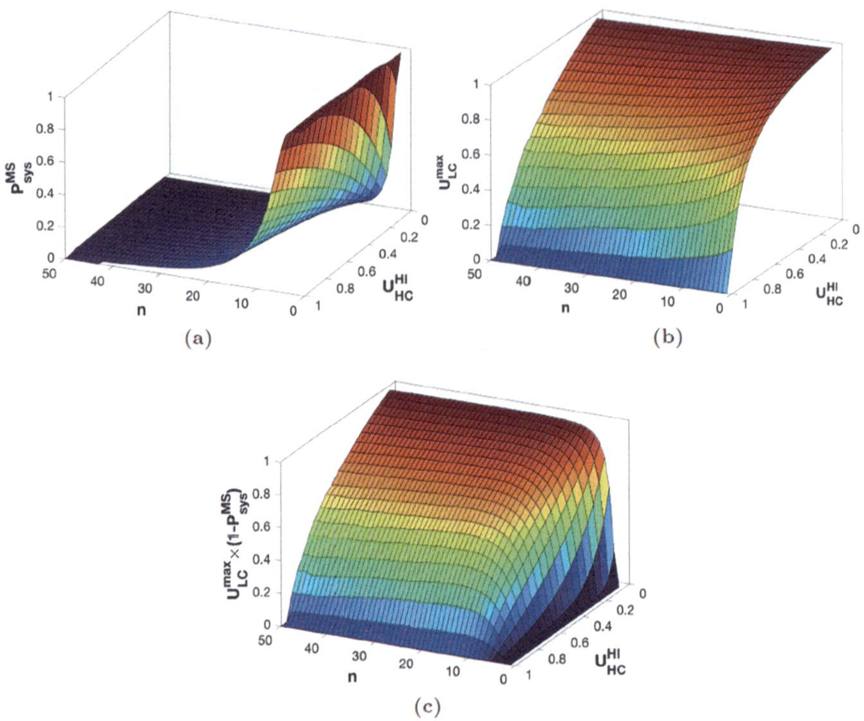

Fig. 3.5 Effect of n and HC tasks' utilization on maximum assigned utilization to LC tasks and mode switching probability. (**a**) P_{sys}^{MS} by varying n and U_{HC}^{HI}. (**b**) $max(U_{LC}^{LO})$ by varying n and U_{HC}^{HI}. (**c**) Optimization goal

task sets using the GA, for different utilization. In Baruah's approach [1] (Bar+12a), considering a large lower-bound value for λ like $1, \frac{1}{2}$ reduces the probability of mode switching, but it underutilizes the system during run-time. For example if $U_{HC}^{HI} = 0.75$, for $\lambda = 1$, $P_{sys}^{MS} = 9.66\%$ and $max(U_{LC}^{LO}) = 23.28\%$, while for our proposed scheme, $P_{sys}^{MS} = 16.46\%$ and $max(U_{LC}^{LO}) = 52.39\%$. On the other hand, using a smaller lower-bound value for λ like $\frac{1}{8}$ increases the maximum utilization of the LC tasks with high mode switching probability. For instance, if $U_{HC}^{HI} = 0.75$, then $P_{sys}^{MS} = 90.75\%$ and $max(U_{LC}^{LO}) = 70.75\%$. Note that, to prevent the figures from being unclear, we only show the result for the $\lambda \in [\frac{1}{8}, 1]$. The results for $\lambda \in [\frac{1}{16}, 1]$ and $\lambda \in [\frac{1}{32}, 1]$ have more maximum utilization increment of the LC tasks with higher mode switching probability in comparison with $\lambda \in [\frac{1}{8}, 1]$, which is undesirable. Our approach works well in both system properties by determining the best $WCET^{LO}$ values for HC tasks base on the ACET and then the optimum U_{HC}^{LO}. Figure 3.6c shows this fact by optimizing both system properties, where the proposed scheme performs better than other policies. As a result, our scheme

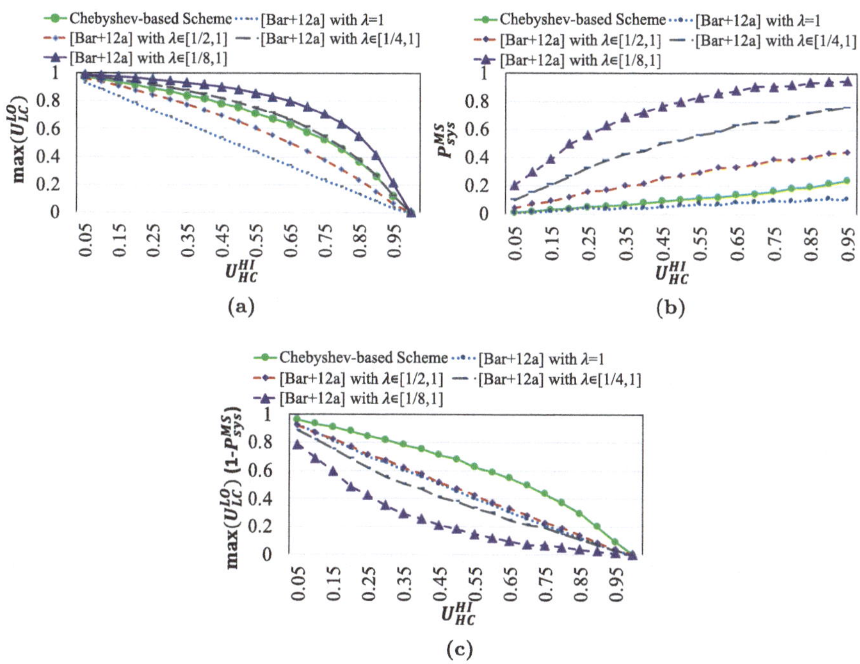

Fig. 3.6 The effectiveness of our proposed scheme in comparison with other policies, proposed in other research works. (**a**) $max(U_{LC}^{LO})$ by varying U_{HC}^{HI}. (**b**) P_{sys}^{MS} by varying U_{HC}^{HI}. (**c**) Optimization goal by varying U_{HC}^{HI}

improves the utilization by up to 72.27% compared to the existing approaches, while P_{sys}^{MS} is bounded by 24.28% in the worst-case scenario.

Evaluating Scheduling Approaches Under the Proposed Scheme Now, we evaluate and compare the results in terms of schedulable task sets (acceptance ratio) to the state-of-the-art approaches, proposed in [1, 7], with and without our scheme. In this experiment, we assume that the probability that a task is an HC or LC is equal. In both [1, 7], the EDF-VD algorithm has been used to schedule the tasks. In [7], the algorithm executes all LC tasks in the HI mode by reducing their WCET to 50%, and also in [1], the algorithm drops all LC tasks when the system switches to the HI mode. It is noteworthy to mention that our scheme for selecting the suitable $WCET^{LO}$ for HC tasks can be applied to any scheduling algorithm with any policy of task execution and optimize the resource utilization and mode switching probability.

Figure 3.7 shows the acceptance ratio for two state-of-the-art scheduling approaches [1, 7], which are improved with our scheme in all utilization bounds. As shown in this figure, when $U_{bound} \leq 0.7$, all task sets are schedulable with Liu's approach [7] and our scheme. When the system utilization is increased ($0.7 < U_{bound} \leq 0.95$), our proposed scheme performs better than Liu's

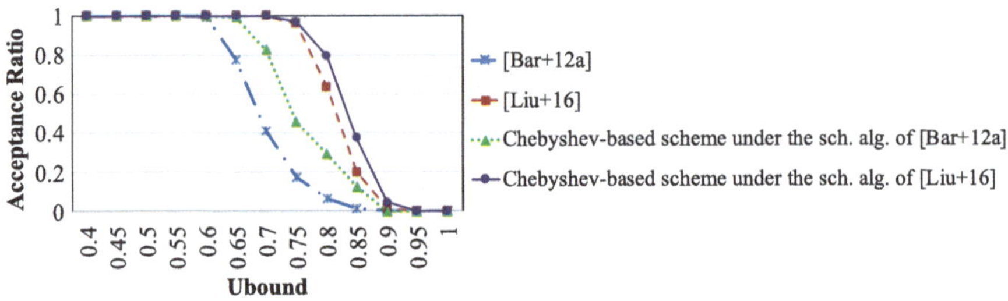

Fig. 3.7 Different scheduling approaches ([1] (Bar+12a) and [7] (Liu+16)) with our scheme

approach [7] in terms of acceptance ratio. And so that no task set is schedulable for $U_{bound} \geq 0.95$. Besides, the same trend is found for Baruah's approach [1]. The reason for having a better acceptance ratio in our scheme is determining the appropriate $WCET^{LO}$ for HC tasks and executing more tasks in the system.

Although this design-time approach (*BOT-MICS*) can improve the QoS and reduce the mode switching probability in comparison with state-of-the-art works, the constant WCETs are set for tasks to be used in the LO mode, which remain unchanged during run-time. Such static techniques cannot employ the benefits of dynamic execution time changes and, therefore, may cause significant performance loss for LC tasks or processor underutilization at run-time if the low WCETs are not close to AETs. Therefore, we propose *ADAPTIVE* in the next section, which is a run-time approach to adapt to task execution time dynamism at run-time and adjust low WCETs.

3.2 ADAPTIVE: A Run-Time WCET Adjustment Approach

In this work, we propose a novel learning-based run-time scheme for determining low WCET ($WCET^{LO}$) to (1) effectively reduce the system switches to the HI mode, (2) have high processor utilization and consequently, a high value of QoS, (3) guarantee the system to be schedulable in each criticality level, and (4) not be affected and varied by sudden changes of execution times. To the best of our knowledge, there is no method so far to determine the low WCETs for MC tasks at run-time based on the behavioral system changes while making a trade-off between the QoS, utilization, and mode switches. The main contributions of this work are:

- Presenting a novel adaptive scheme to analyze and obtain the low WCETs of MC tasks at run-time and manage the mode switching probability and QoS
- Proposing a learning-based mechanism, called *ADAPTIVE*, to improve the MC system timing behavior at run-time
- Presenting a dynamic QoS-aware scheduling algorithm to improve the results' quality at run-time based on the system changes while guaranteeing the minimum service of LC tasks, even in the HI mode

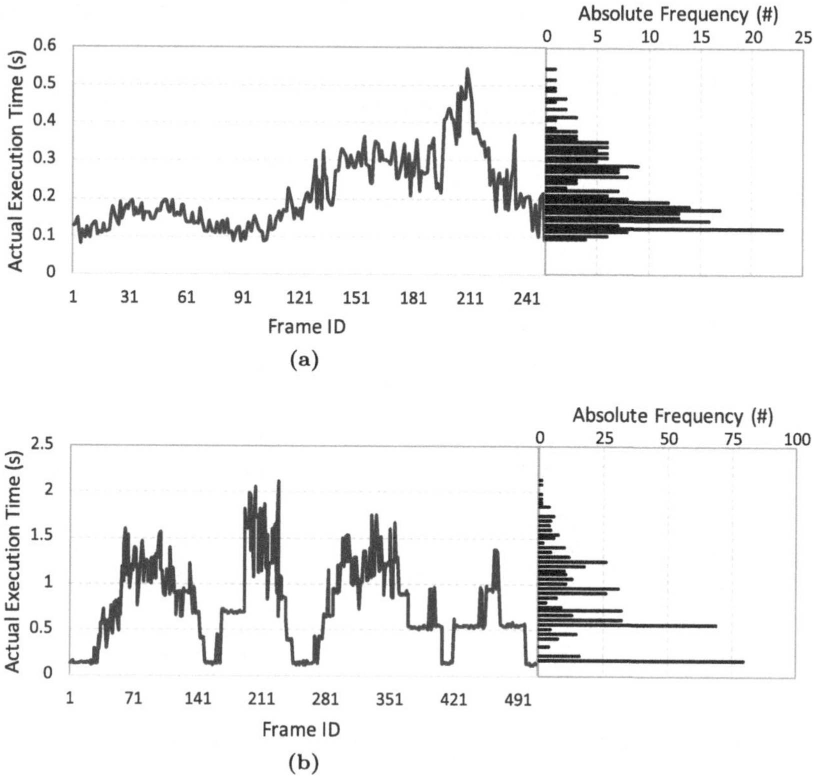

Fig. 3.8 Execution time values for two different time recording videos as input for object detection function during run-time and their time distribution. This figure shows that both aspects of run-time and design-time behavior should be considered in MC system design and task property determination. (**a**) Input video with few objects to detect. (**b**) Input video with few and many objects to detect

3.2.1 Motivational Example

In general, the actual execution time of tasks depends on their input values. Due to the spatial or temporal correlation in the input data stream like video, the execution times of the tasks are often temporally correlated. Figure 3.8 shows the computational times of the object detection function running on the ODROID XU4 board powered by ARM Cortex A7. Note that the object detection function is one of the main functions in an autonomous driving application—an MC system. For input, videos from a road camera in the two different time slots, converted to motion jpegs, are given to the function of detecting cars on the road. The videos were recorded for a period of time when it experienced both light and heavy traffic. Figure 3.8 shows how the computation times of detecting objects vary during run-time. The computation time values in this function depend on the number of objects to be detected. As we can see, the times of multiple jpeg images are clustered due to the temporal correlation between the subsequent inputs presented to the application. For this example, static approaches such as the one presented in our previous work [23, 24] and [2, 15] set the static $WCET^{LO}$, considering the execution time of the majority of instances. This static WCET works fine for some time, but it may lead to frequent mode switches when there are many objects to detect (e.g., heavy traffic) or poor utilization when there are few objects to detect in this function (e.g., light traffic). As a result, proposing an adaptive scheme to determine the low WCETs dynamically may significantly improve the mode switches, QoS, and utilization. Therefore, the system's run-time behavior can be investigated by monitoring the AETs and adjusting the low WCETs.

3.2.2 *ADAPTIVE in Detail*

The goal of our proposed scheme (*ADAPTIVE*) is to improve the LC tasks' QoS as the system utilization while reducing the number of mode switches (MS_{HC}) at run-time. The values of $WCET^{LO}$s for HC tasks have a key role in improving the system objectives. Therefore, it is a challenge to set $WCET^{LO}$ for each HC task to draw a trade-off between the objectives: utilization of the system and the number of mode switches. To address the challenge, we monitor the run-time execution times of HC tasks and adapt $WCET^{LO}$ of HC tasks at run-time to achieve a higher system's QoS, while having fewer mode switches, based on the variation in execution times due to the input and environmental changes. Figure 3.9 shows an overview of the proposed approach (*ADAPTIVE*), which consists of design-time and run-time phases. Here, the task schedulability must be guaranteed at design-time and run-time, and the low WCET adaptation is done at run-time. We further explain them in detail in their corresponding sections.

3.2.2.1 Design-Time Exploration

In order to analyze and schedule the HC and LC tasks in the system, first, the WCETs, required by the tasks, must be obtained. Here, the $WCET^{HI}$ (which is used in the HI mode) of HC tasks are computed by using the OTAWA tool [5], which provides a safe and conservative execution time bound. The WCETs of LC tasks can also be determined by using the OTAWA. In addition, to obtain the $WCET^{LO}$, we run the benchmarks with various data set inputs on ARM Cortex A7, in ODROID XU4 board, and set the maximum value of these actual execution times, as $WCET^{LO}$ for each HC task. Since a periodic task model is considered in this work, the periods are the system inputs at design-time. Since multiple tasks are executed in the system and tasks have distinct periods (not the same period values), the hyper-period, which is the Least Common Multiple (LCM) of all tasks' periods, is used while analyzing the system and task schedulability. The hyper-

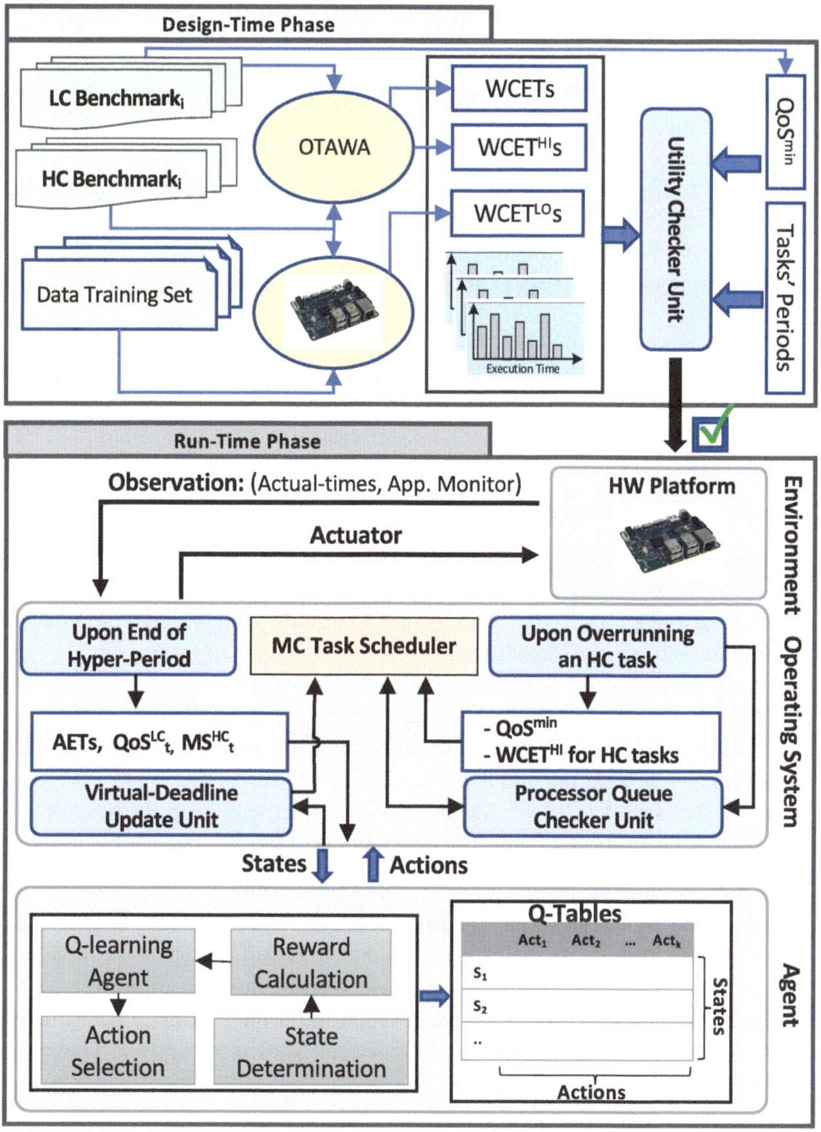

Fig. 3.9 An overview of design-time and run-time phases in ADAPTIVE

period represents a time that there is no workload in a system after frequent task releases and scheduling. These inputs and analyzed data have been used to check the task schedulability by the Utility Checker Unit, which is shown in the design-time phase of Fig. 3.9.

In this work, the EDF-VD scheduling algorithm [2] is applied to schedule MC tasks. However, the proposed scheme is applicable to any scheduling algorithm. If

U_l^k denotes total utilization of tasks with the same criticality level l ($l \in \{LC, HC\}$) in the mode k ($k \in \{LO, HI\}$), where $U_l^k = \sum_{i \in \{LC, HC\}} \frac{WCET_i^k}{T_i}$, Eq. (3.20) must be satisfied to guarantee schedulability by EDF-VD. This equation presents the necessary and sufficient conditions to guarantee the task schedulability in both LO mode and HI mode and meeting the deadlines, even if the system switches to the HI mode [2, 23]. Although the maximum value of utilization is desired, as can be seen from Eq. (3.20), it is upper-bounded by the schedulability constraints. The utilization (UMC) which is the maximum value of two phrases shown in Eq. (3.20) must be less than one in EDF-VD at all time, which is checked by Utility Checker Unit:

$$U_{HC}^{LO} + U_{LC}^{LO} \leq 1 \quad \& \quad U_{HC}^{HI} + U_{LC}^{HI} + \frac{U_{HC}^{LO} \times (U_{LC}^{LO} - U_{LC}^{HI})}{1 - U_{LC}^{LO}} \leq 1 \tag{3.20}$$

3.2.2.2 Run-Time Adaptation

The crucial research questions that should be addressed in the run-time phase are:

1. How to vary WCET of HC tasks in the LO mode with no adverse effect on meeting the other tasks' deadlines
2. How the scheme should be designed for determining the $WCET^{LO}$ at run-time, to not be affected and varied by sudden changes in execution times
3. How to design a scheme with low timing overheads during run-time to have no impact on task scheduling and deadline misses
4. What are the best $WCET^{LO}$ for the tasks to effectively keep the system away from switching to the HI mode while having the high processor utilization and consequently, a high value of QoS

Following the above questions, the ML techniques can effectively help to design an adaptive MC system to make a reasonable trade-off between the objectives according to the system environmental changes (i.e., input value variation).

At run-time, the MC tasks start their execution on the platform, controlled by MC Task Scheduler Unit on the operating system, shown in Fig. 3.9. The system monitors the tasks from two aspects:

1. Each task execution finishes or not: The actual execution times are stored in the case of complete execution. In addition, in the case of task overrunning, the system switches to the HI mode, and the MC Task Scheduler Unit executes the HC tasks by considering their $WCET^{HI}$ and LC tasks with their QoS^{min}. The Processor Queue Checker Unit keeps track of the processor queue when the system can switch back to the LO mode.
2. The system reaches the task set hyper-period or not: At the end of each hyper-period, the agent starts its operation by employing the data like actual execution times, the number of mode switches, and the QoS of LC tasks in the last hyper-

period. The agent outputs are the new values of $WCET^{LO}$ for HC tasks, used in the next hyper-period, based on these historical data. Since the utilization of HC tasks in the LO mode would be changed by updating $WCET^{LO}$s, the new virtual deadlines are determined by the Virtual-Deadline Update Unit.

Hence, the learning process is separate from the task scheduling algorithm, and we do not use learning techniques to schedule the tasks. The EDF-VD schedulability formulae are checked for each WCET change (at the end of each hyper-period). Although this time is in the order of microseconds and can be negligible, we counted this time as part of learning time. The timing overhead of this process is considered a task with the WCET, equal to the maximum timing overhead to ensure it does not impact other tasks' deadlines. This overhead is discussed and reported in Sect. 3.2.3.3. In the following, we describe how the agent is designed to work and update the WCETs for the LO mode.

Learning-Based System Properties' Improvement Reinforcement Learning (RL) could be applied to systems with a considerable amount of dynamism through trial and error. By using the historical data, and learning from past events, it is able to improve the performance, based on the dynamic changes [25]. The Q-learning/SARSA technique, which is recently used in many emerging applications, such as robotics, and UAV [26], uses the RL technique to perform the run-time management/optimization of the system properties. This technique is a value-based algorithm that iteratively collects the current system state and determines the next action to change the state. The process repeats until the predefined criterion is met or the objectives are no longer significantly improved.

RL technique consists of the three main components: (1) a discrete set of States $= \{state_1, state_2, \ldots, state_l\}$, (2) a discrete set of Actions $= \{Act_1, \ldots, Act_k\}$, and (3) reward function $Reward$ [25]. To reach the favorable reward, the technique learns a lookup table (i.e., Q-table) with $(state_t, Act_t)$ pairs ($Act_t \in Actions$ and $state_t \in States$). The states and actions determine the rows and columns of the Q-table of the learning-based algorithm, respectively (shown in Fig. 3.9). As mentioned, a value-based algorithm is utilized which is represented with $Q(state_t, Act_t)$ in the Q-table and determines the quality of the taken action at the particular state. In every iteration, the Q-values are updated based on the corresponding computed reward according to Eq. (3.21), which is based on the SARSA learning algorithm, one of the RL methods, for system objective improvement [27, 28]:

$$Q(state_t, Act_t) = Q(state_t, Act_t) + \alpha(Reward^{MC} + \gamma Q(State_{t+1}, Act_{t+1}) - Q(State_t, Act_t)) \quad (3.21)$$

where $state_t$ and Act_t represent the state and action of the system at time t, respectively. Furthermore, $state_{t+1}$ and Act_{t+1} indicate their values at time $t+1$. The α determines the learning rate of overriding the old data in the table by the new acquired data ($0 < \alpha \leq 1$). $Reward^{MC}$ is the reward function, and γ is the discount rate to determine the importance of the future reward ($0 < \gamma < 1$).

System State Determination There are various criteria for determining the system states. In *ADAPTIVE*, the system states (i.e., the rows of the Q-table) indicate the rate of LC tasks' execution, i.e., the tasks' periods at run-time, to the minimum tolerable period in LC tasks. We define ten ranges to determine the rate of LC tasks' execution. As a result, $State_s = \{0.1, 0.2, \ldots, 1\}$.

Learning Action Determination In this section, the well-known ϵ-greedy policy (a method for determining the optimal action according to the state) has been exploited. In this policy, a random action is selected from the actions set with the probability of ϵ. In general, a random number is generated. If this number is less than ϵ, a random action is selected from the action set; else, the best action is selected with the largest Q-value (which can be with the probability of $1 - \epsilon$). We first use a dynamic ϵ-greedy policy [29] with the maximum value of 0.5 to prevent the probability of the learning algorithm from being stuck at a few Q values. Afterward, the fixed ϵ-greedy policy is used to ensure that the system reaches the optimum state and chooses the best action based on the Q-values, which has the maximum value. We have assumed k actions, where the action space in the Q-table illustrates an increase and/or decrease in the $WCET_i^{LO}$ of some/all HC tasks according to a coefficient of WCET's prediction accuracy. In order to limit the number of feasible actions and reduce the complexity and convergence issues, we have considered three scenarios of increase ($WCET_{NumT}^{inc}$), decrease ($WCET_{NumT}^{dec}$), and increase-decrease ($WCET_{NumT}^{inc,dec}$) ($opr = \{inc, dec, inc/dec\}$). $WCET_{NumT}^{inc}$ ($WCET_{NumT}^{dec}$) shows an increase (decrease) in $WCET_i^{LO}$ of $NumT$ HC tasks, where the value of $NumT$ can be one of $\{1, 2, \ldots, n_{HC}\}$ (n_{HC} is the maximum number of HC tasks in the system). $WCET_{NumT}^{inc/dec}$ presents that the $WCET_i^{LO}$ for half of HC tasks is increased and for others it is decreased. In fact, in this scenario, $max(NumT) = \frac{n_{HC}}{2}$. Therefore, we have considered $k = 2.5 \times n_{HC}$ actions in the system. Note that the step of increase/decrease in the $WCET_i^{LO}$ is determined according to a coefficient of WCET's prediction accuracy:

$$Actions = \{WCET_{NumT}^{opr}\} \quad opr \in \{inc, dec, inc/dec\} \tag{3.22}$$

To select the tasks for doing the actions, we first sort the tasks in increasing order of the value of $WCET_i^{LO} - AET_i$ in the last hyper-period, and then the increased (decreased) action applies to the $NumT$ tasks with smaller (greater) $WCET_i^{LO} - AET_i$, where AET_i represents the actual execution time. Since a task may release several times in a hyper-period (i.e., release several jobs of a task), and the actual execution times would be different in each release time, we have to predict the actual execution time according to the previous run-time task's execution times. This prediction is based on the following equation, where $ExeTime_i(t + 1)$ is the predicted execution time of task τ_i for hyper-period HP_t, rc_i is the regression coefficient, and er is the error (presents how different the estimated value is from the actual one). In the evaluations, x is assumed to be 8. This number is chosen based on various experiments that we performed to achieve lower $ExeTime$ prediction error with no timing overhead that can impact tasks' timeliness. For example, for one task, we have (x,er,time[$\mu Second$])=(2,0.110,0.86), (4,0.094,1.12), (8,0.077,1.59), (10,0.071,1.92). Since the error (er) does not change much, from 8 to 10, compared to 4 to 8, x= 8 is a good value with less timing overhead:

$$ExeTime_i(t+1) = \sum_{k=0}^{x} ExeTime_i(t-k) \times rc_k + er \tag{3.23}$$

Reward Computation The reward indicates how well the learning procedure has performed in the previous step. In *ADAPTIVE*, we calculate the reward at the end of each hyper-period. Here, the number of mode switches should be reduced while increasing the number of scheduled LC tasks to improve the QoS. The considered reward function for the Q-table is based on these two objectives, shown in Eq. (3.24), where MS_{HC} represents the mode switches, which is determined by the number of overrun HC tasks:

$$Reward^{MC} = \beta_1 \times MS_{HC} + \beta_2 \times QoS \tag{3.24}$$

where β_1 and β_2 are constants in Eq. (3.24) and set to 0.5 ($\beta_1 + \beta_2 = 1$) in this work. To compute the number of mode switches, Eq. (3.25) considers three scenarios. The reward function is calculated based on the number of task overruns. If the percentage of overrun HC tasks falls into the unsafe zone that may cause frequent mode switches, the decision will be penalized. Accordingly, it results in a negative value for the reward function, which decreases the Q-value in Eq. (3.21). If the number of task overruns is reduced, the higher and positive value is assigned to MS_{HC}. The unsafe zone is the statement that all HC tasks overrun frequently. Equation (3.26) also shows how to compute the percentage of overrun HC tasks:

$$MS_{HC} = \begin{cases} +\Gamma & PT_{HC}^{ovr} = 0 \\ 1 - \frac{1}{10\times(1-PT_{MC}^{ovr})} & 0 < PT_{HC}^{ovr} < 1 \\ -\Gamma & PT_{HC}^{ovr} = 1 \end{cases} \tag{3.25}$$

where $\Gamma > 0$ and has a constant value and is set to 1 in this work:

$$PT_{HC}^{ovr} = \frac{\#HC - Tasks|_{\frac{WCET_i^{LO} - ExeTime_i}{T_i} < 0}}{n_{HC}} \tag{3.26}$$

n_{HC} is the number of HC tasks and $ExeTime_i$ is the estimation of actual execution time of task τ_i during a hyper-period, which is computed by Eq. (3.23).

Algorithm 3.1 Run-time adaptation scheme

Input: *Task Set, Single Processor Platform,* QoS^{Min}
Output: QoS, $WCET^{LO}_{HC}$s, *Scheduled Tasks*
1: **procedure** ADAPTIVE FUNCTION()
2:   **for** t = 1 **to** *Time* **do**
3:     $[Mode^{Sys}_{MS}, ReadyTask]$ = **TaskStatusCheck**(*Tasks*,$Mode^{Sys}_{MS}$)
4:     $[Sch_{tasks}]$ = **EDF-VD** ($ReadyTask$, *Platform*)
5:     $Flag_{output}$=**TaskOutputCheck**(*Tasks*)
6:     **if** $Flag_{output} == 1$ **then**
7:       Update QoS & PT^{ovr}_{HC};
8:     **end if**
9:     //**Learning Process Function**
10:     **if** mod(t,HP)==0 **then**
11:       *State*= **Deter-State** ($\#Scheduled - LCTasks$)
12:       *k= rand (1)*; //$(0 < k < 1)$
13:       //ϵ*-Greedy Policy*
14:       **if** $k < \epsilon$ **then**
15:         Act_t = *argrand (Actions)*
16:       **else**
17:         $Act_t = argmax\ (state_t,$ Actions)
18:       **end if**
19:       *Set the new tasks'* $WCET^{LO}$*s based on the action*
20:       $Reward^{MC}$ = **CompRward** (PT^{ovr}_{HC}, QoS) //Eq. (3.24)
21:       $Q(state_t, Act_t) = Q(state_t, Act_t) + \alpha(Reward^{MC} + \gamma Q(state_{t+1}, Act_{t+1}) - Q(state_t, Act_t))$ //Eq. (3.21)
22:       $UMC(t)$=CompUtil ($Tasks$)
23:       **if** $UMC(t) > 1$ **then**
24:         $Tasks$ = Deter-ExeJobs ($Tasks$);
25:       **end if**
26:       $Tasks$ = Deter-VirtualDeadline ($Tasks$);
27:     **end if**
28:   **end for**
29: **end procedure**

Algorithm The pseudo-code for the run-time adaptation scheme of *ADAPTIVE*, which includes the task scheduling and learning procedures, is presented in Algorithm 3.1. As inputs, the algorithm takes the tasks, and their characteristics (such as WCETs), platform, and the minimum QoS, requested by the tasks. On the other hand, improvements of the LC tasks' QoS, the analyzed $WCET^{LO}$s of HC tasks, and the scheduled tasks are defined as outputs at the end of time ($Time$). At each time, the scheduler checks the status of the tasks, whether they are released or overrun, which results in mode switching (line 3). All tasks are scheduled based on the EDF-VD (line 4). In the case of the system switching to the HI mode, the LC tasks must be executed based on their minimum service requirement to guarantee the correct execution of HC tasks. There is a function (line 5) that checks whether the output of each task is ready. In the case of being ready, the task is removed from the core queue, and the values of QoS and PT^{ovr}_{HC} are updated (lines 6–8). The learning process is conducted at the end of each hyper period (lines 9–27). The number of scheduled LC tasks is used to determine the state (line 11) in this process. As mentioned earlier, since the ϵ−greedy policy has been used, if a random number is less than ϵ, a random action is selected (line 15, exploration phase of the learning process); otherwise, the action that has the maximum value in the Q-table is selected for that particular state (line 17, exploitation phase of learning process). Based on the chosen action, new $WCET^{LO}$ values are determined for some HC tasks (line 19). Consequently, the reward function is used to update the Q-table based on Eq. (3.24) (lines 20–21). In lines 22–26, the guaranteed service adaptation and assigned virtual deadlines to HC tasks are computed based on the updated HC tasks' utilization values to guarantee the system to be schedulable in each criticality level. As a result, the maximum service adaptation that can be guaranteed is determined by finding the maximum value of rate of LC tasks' execution, i.e., reducing the LC tasks' periods to release more often.

3.2.3 *Evaluation*

In this section, we evaluate the efficacy of *ADAPTIVE*, on real-life and synthetic task sets in terms of mode switches, QoS, utilization waste, and learning process timing and memory overheads.

3.2.3.1 Evaluation with Real-Life Benchmarks

To evaluate our scheme, we conducted experiments by various real benchmarks from MiBench benchmark suite [12], such as automotive, network, and telecommunication. In this experiment, ‹insert-sort›, ‹epic›, ‹qsort›, ‹bitcount›, ‹dijkstra›, and ‹FFT› are considered as HC tasks and ‹corner›, ‹edge›, ‹smooth›, and ‹matrix-mult› are considered as LC tasks. To obtain their execution times, we run these benchmarks with different inputs on Cortex A7 of the ODROID XU4 board (equipped with Ubuntu 18.04 as OS) with a maximum frequency of $1.4GHz$. More detail on WCET values has been reported in Sect. 3.1.3. We compare the results with the results of *BOT-MICS* [24] (*the Chebyshev theorem*-based one) and [2]. Since most state-of-the-art works like [2, 3, 15] consider the same policy to determine the $WCET^{LO}$ (i.e., defining a fraction of $WCET^{HI}$ as $WCET^{LO}$), we select [2] ([Liu+18]) as a representative of these schemes and do the experiments with two fractions of $WCET^{HI}$ as $WCET^{LO}$ ($\lambda = \frac{WCET^{LO}}{WCET^{HI}} = [\frac{1}{4}, 1], [\frac{1}{8}, 1]$). In addition, we investigate the system for 2000 hyper-periods of tasks.

For the learning process, we set the values of γ to 0.2 and α to 0.5. These values are determined based on a wide range of experiments, which are set to obtain the best improvement.

Table 3.6 presents the evaluation of different approaches. QoS represents the percentage of executed LC task instances to total LC task instances during run-

Table 3.6 System performance at run-time for different scenarios

	$Avg(QoS)$[a]	$Avg(\#MS^{HC})$[b]	$Util^{Wst}$[b]	$max(U_{LC}^{LO})$[a]
[2] $\lambda = \frac{1}{4}$	50.0%	0	43%	50%
[2] $\lambda = \frac{1}{8}$	49.3%	5.81	28%	65%
BOT-MICS	58.1%	1.16	33%	63%
ADAPTIVE	68.9%	2.12	16%	58%

[a] Higher is better
[b] Lower is better

Table 3.7 Number of deadline misses and gained utilization of different methods at run-time for object detection function in Fig. 3.11a, where there are many objects to detect

Metrics	*ADAPTIVE*	BOT-MICS	[2] $\lambda = \frac{1}{2}$	[2] $\lambda = \frac{1}{4}$	[2] $\lambda = \frac{1}{8}$
$\#MS^{HC}$	17%	11%	0	5%	45%
$Util^{Wst}$	28%	46%	76%	52%	47%

time ($QoS = n_{LC}^{schd}/n_{LC}^{max}$). MS^{HC} indicates the number of mode switches per hyper-period, and $max(U_{LO}^{LC})$ represents the maximum processor utilization that can be assigned to LC tasks at design-time. Besides, $Util^{Wst}$ shows the average percentage of the difference between the WCET and AETs to WCET for all tasks. As shown in this table, *ADAPTIVE* scheme can schedule more LC tasks in both LO mode and HI mode, 10.8% and 19.3% improvement, compared to *BOT-MICS*, and [2] approaches, respectively. Although the average number of mode switches is more than the scenario of *BOT-MICS*, and [2] with $\lambda = \frac{1}{4}$ (hence, the LC tasks execute with their minimum service requirements in the HI mode), *ADAPTIVE* could overcome the significant performance loss due to executing more LC task instances in total. This can be achieved by determining the appropriate $WCET^{LO}$s during run-time, which is close to the actual execution times (17.7% closer on average, compared to other approaches). This fact can be observed with the value of $Util^{Wst}$, compared to other approaches. In addition, although the scenario of [2] with $\lambda = \frac{1}{8}$ assigns more utilization to LC tasks in the design-time phase compared to other approaches, it has less QoS value due to the more frequent mode switches. Note that although we assign a primary value of $WCET^{LO}$ by running each benchmark on the platform and set the maximum of them as $WCET^{LO}$ (which causes lower utilization compared to some other approaches), our run-time learning approach is independent of how the $WCET^{LO}$s are set at design-time, and any of other design-time approaches, like our previous approach, *BOT-MICS*, can be used.

Figure 3.10 shows the variation of QoS values in different hyper-periods of the run-time phase. As shown, the scenario of [2] with $\lambda = \frac{1}{8}$ has a wide range of QoS values due to more mode switches. *ADAPTIVE* also represents a variation in the QoS values due to its adaptation to the system input changes at run-time.

Now, we demonstrate the progress of the learning process in adjusting the $WCET_i^{LO}$ for the two input videos from the object detection function, explained in

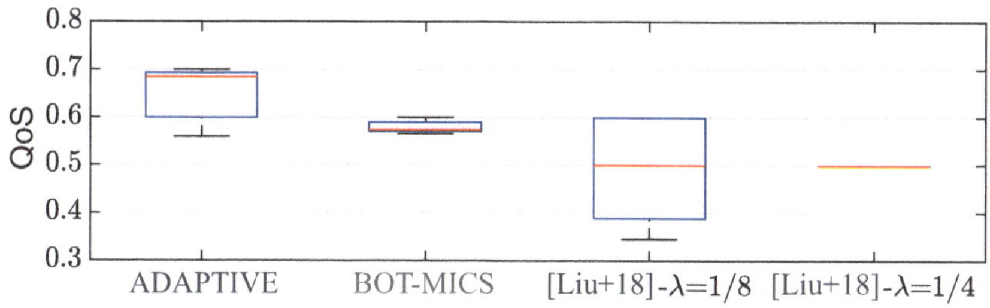

Fig. 3.10 QoS values during run-time for different scenarios (ADAPTIVE, BOT-MICS [24], and [Liu+18] [2])

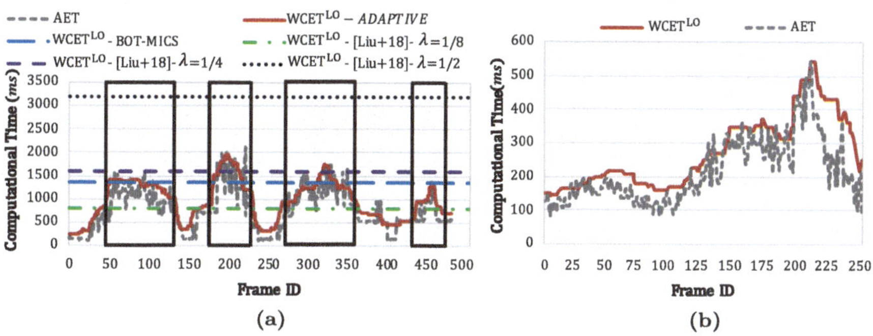

Fig. 3.11 Learning process in adjusting the $WCET^{LO}$ for two video inputs of object detection function, compared to state-of-the-art method ([2] (Liu+18)) in adjusting the $WCET^{LO}$ value. (**a**) Input video with many objects to detect. (**b**) Input video with few objects to detect

the motivational example of the Introduction Section, during run-time. Figure 3.11 depicts the actual execution time trace and the adjusted $WCET^{LO}$ for a time period during run-time for *ADAPTIVE* and two other methods *BOT-MICS* [24] and [2]. In *ADAPTIVE*, by changing the inputs which have low execution time values, the $WCET^{LO}$ would be adjusted to the lower value intentionally. The $WCET^{LO}$ is also readjusted when the value of execution times is increased. As shown in Fig. 3.11a, the other methods presented in *BOT-MICS* and [2] set the static $WCET^{LO}$ (as an example, $WCET^{LO}$=1352 ms in *BOT-MICS* approach and $WCET^{LO}$ ($WCET^{LO}$=1596 ms in [2] approach if $\lambda = 1/4$) which may lead to frequent task overruns (which leads to regular mode switches) in the case that there are many objects to detect (lead to high computational time values) or poor utilization when there are few objects to detect in this function (like Fig. 3.11b, where the maximum computational time is 548 ms in a period of time shown in this example). Although there are few errors while adjusting the $WCET^{LO}$, which leads to task overrunning and QoS degradation, the number of task overruns for the HC task and the wasted processor utilization ($Util^{Wst}$) are less, and consequently, overall QoS value would be higher at the end of system execution. For example, in Fig. 3.11a, the black rectangles show the time periods in which some of the

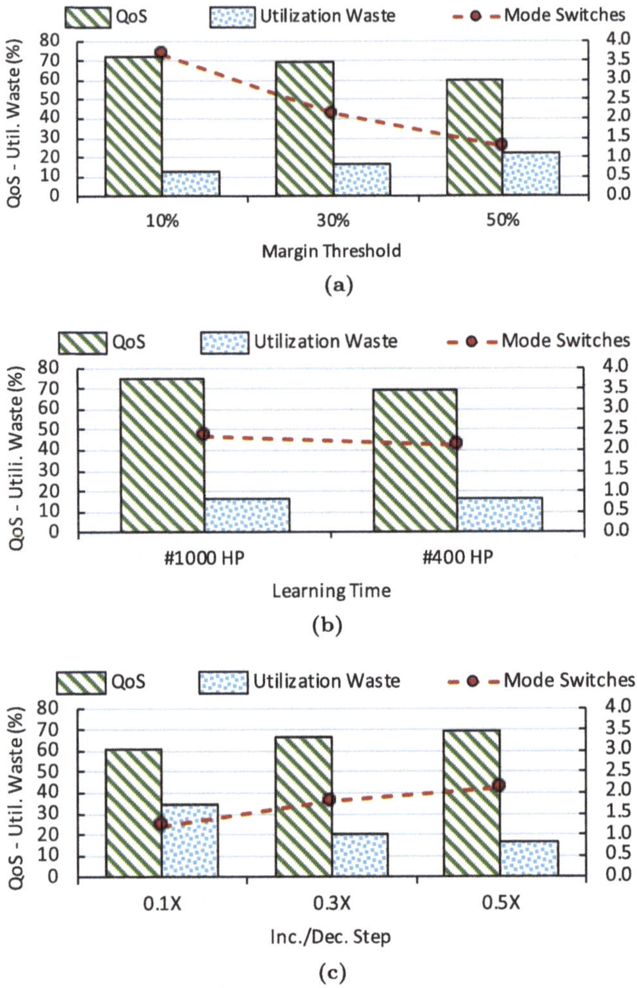

Fig. 3.12 Impacts of varying different parameters of the learning process on QoS, mode switches, and utilization waste. (**a**) Varying margin threshold. (**b**) Varying learning time. (**c**) Varying WCET increase/decrease steps

actual execution times are higher than $WCET^{LO}$ for most methods, even though the actual execution times of these input videos are less than the adjusted $WCET^{LO}$ in [2] by considering $\lambda = \frac{1}{2}$. Table 3.7 presents the percentage of deadline misses (which leads to mode switches) and the average percentage of wasted utilization for *ADAPTIVE* and state-of-the-art works. As shown, although the number of deadline misses is higher than the results of some scenarios, the wasted utilization is lower compared to other methods, which leads to higher QoS (QoS is like what is discussed in Fig. 3.10 and Table 3.6 for MiBench benchmarks).

In order to evaluate *ADAPTIVE* in improving the QoS, we first analyze varying the margin threshold, which adjusts above the actual execution times to overcome high WCET reduction. For example, when the threshold is x%, $WCET^{LO}$ is set to (1 + x%) of the maximum estimated execution time. Figure 3.12a shows that the improvement in QoS is less while the threshold is pessimistic, i.e., having a larger margin. In this case, fewer mode switches exist due to adjusting the WCETs so cautiously, and consequently, the utilization waste ($Util^{Wst}$) has a higher value. Based on this observation, having the margin threshold equal to 10% improves the QoS for both LO mode and HI mode, even with the higher number of mode switches. Now, we vary the time of learning during run-time, e.g., 400 or 1000 Hyper Period (HP) of total time (2000 HP), which is used for training/exploring and depicted in Fig. 3.12b. Note that the learning process starts to learn for a period of time, and then the learned data is exploited for the rest of the time. Spending more time to learn leads to more accurate results. As a result, we would have a 5.9% improvement in QoS by increasing the learning time, with an insignificant increase in mode switches and utilization waste. In order to be more accurate in adjusting the WCETs, we define steps for increasing/decreasing the WCETs at run-time. These steps are coefficients of the difference between the WCETs and actual execution times. Having a larger coefficient (0.1X to 0.5X) leads to adjusting faster to the actual execution times (i.e., having better QoS values (7.95% more) and less utilization waste (18%)), but it may cause more wrong decisions in learning (i.e., more mode switches). However, step $= 0.5X$ improves the QoS more, which is computed for both LO mode and HI mode.

3.2.3.2 Evaluation with Synthetic Task Sets

We now carry out an extensive evaluation with the synthetic task sets to evaluate the proposed scheme effectiveness, compared to our previous proposed approach (*BOT-MICS* Chebyshev theorem-based one), and the state-of-the-art work [2] in terms of varying utilization in Fig. 3.13. The synthetic task sets are generated for various utilization bounds (U_{bound}) in line with research works like [3], in the range of [0.60,1] with steps of 0.05. For each U_{bound}, 50 task sets are generated in which tasks' periods are selected in the range of [200, 1000] ms. The algorithm adds tasks to the task set randomly to increase the U_{bound} until it reaches a given threshold. Besides, the balanced tasks are assumed in terms of criticality levels, i.e., the probability of a task being HC or LC is equal. As one of the algorithm inputs, we consider the minimum service requirement of LC tasks equal to 0.3 in this section. According to Fig. 3.13, the ability to improve the QoS is less by increasing the utilization bound due to having more HC and LC tasks in the system, and then, the probability of mode switches is increased. In [2], although considering a small value for λ like $\frac{1}{8}$, increases the assigned utilization to LC tasks, it causes more mode switches and dropping more LC task instances, which leads to poor QoS. Besides, considering a large value ($\lambda = \frac{1}{4}$) decreases the utilization at design-time, but it increases the QoS due to a fewer number of mode switches. The *BOT-MICS*

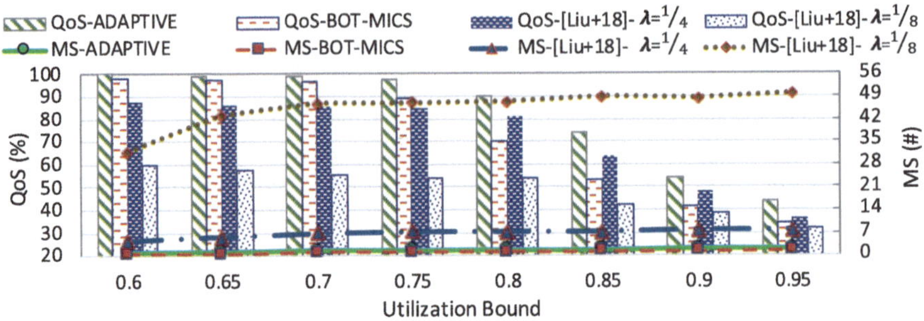

Fig. 3.13 QoS and mode switches in different approaches (ADAPTIVE, BOT-MICS [24], and [Liu+18] [2]) by varying utilization

approach [24] has better results in total in comparison with [2] due to making an ideal trade-off between the mode switches and utilization. However, although *BOT-MICS* has a slight improvement in mode switches, compared to *ADAPTIVE*, our scheme can improve the QoS more, even with more mode switches, due to considering the aspects of run-time behavior, i.e., input and environmental changes, which cause different execution times.

3.2.3.3 Investigating the Timing and Memory Overheads of Learning Technique

Although we have reduced the feasible actions to reduce the complexity and convergence issues, we investigate the learning's timing and memory overheads. We analyze the timing overhead of the learning process in each hyper-period on ODROID XU4, ARM Cortex A7, with 1.4GHz. Consider a system with n tasks, in which n_{HC} of them are HC tasks ($n_{HC} \leq n$). The timing overhead of the learning algorithm is different for the exploration and exploitation phase of the learning process. Hence, we use the ϵ−greedy policy, which makes a trade-off between exploration and exploitation of the learning algorithm. We measured the learning process time at run-time, and the average and maximum of exploration (exploitation) timing overhead in ARM core are 19 µs (52 µs) and 2.11 ms (4.15 ms), respectively. As a result, since the maximum timing overheads are almost significant for real-time systems, we can consider the learning process as a task with the WCET, equal to the maximum learning timing overhead, and a period equal to hyper-period, while checking the task schedulability, at design-time. As a result, it can guarantee that the timeliness of HC tasks is maintained at run-time. Besides, from the memory overhead perspective, we need to clarify the required memory space for storing the Q-table. We store a two-dimensional array with size (*State*) rows and size (*Action*) columns. Since the value of a table cell is in the range of [−2,2], which is a float number, it is required to consider at most 32 bits for storing each cell. As a result, we need size(*State*)×size(*Action*)× 32 bits to store the Q-table. For an application with 40 tasks ($n = 40$), and 20 HC tasks ($n_{HC} = 20$), the amount of required memory space for saving the Q-table with ten states would be $10 \times 2.5 \times (20) \times 32$ bits $= 16$ KB.

3.3 Conclusions

In this chapter, we proposed two novel schemes based on the application analysis to adjust the low WCET of HC tasks in order to improve the QoS of LC tasks. In the first approach, the scheme, called *BOT-MICS*, analyzes the application in the offline phase and determines the low WCETs based on the *Chebyshev theorem*, a general theorem that is valid for any task with any execution time distribution. However, we analyze the applications based on their distribution to have a tighter bound for system mode switching probability. The proposed scheme based on the *Chebyshev theorem* improves the system utilization and schedulability up to 72.27% and 91.2%, respectively, while bounding the mode switching probability to 24.28% in the worst-case scenario. We also evaluated the approaches with real benchmarks on a hardware platform to show their efficacy. The proposed scheme based on the application time distribution analysis can reduce the mode switching probability by 4.85% more for a real task set compared to the scheme based on the *Chebyshev theorem*.

Then, an adaptive scheme, *ADAPTIVE*, was proposed to analyze the HC tasks in the LO mode at run-time to determine their WCET based on the task behavioral changes. The proposed adaptive scheme employed the ML techniques to improve the QoS, i.e., the timing budget allocated to LC tasks, while the task schedulability and timeliness can be guaranteed. The proposed scheme improves the QoS for synthetic and embedded real-time benchmarks by 17.62% and 16.4% on average, respectively. Further, we presented results of the object detection function, commonly used in the automotive domain. We evaluated the function with few and many objects to detect at run-time and saw that the wasted processor utilization is less in *ADAPTIVE* compared to state-of-the-art works. Consequently, the overall QoS value would be higher at the end of system execution.

As proposed in the chapter, two criticality levels of tasks have been considered while designing an MC system. In order to present how the proposed approaches in Sect. 3.1 can be extended to support tasks with multiple criticality levels, consider an MC system which has five sets of tasks in terms of criticality (A, B, C, D, E) in avionic applications (A and E have highest and lowest criticality levels in the system, respectively). In general, for a set of tasks with the lowest criticality level (E), we can choose a small "n" to reduce the "$WCET^{LO}$" value (based on our proposed method) and improve the number of tasks that can be scheduled in the system. Since these tasks have LC levels, dropping them during run-time does not harshly affect the system. Besides, for example, tasks with criticality level C have three WCETs. For the first and second WCETs, we choose a small "n" for computing the first WCET, a bigger "n" for computing the second WCET, and $WCET^{HI}$. When the tasks with criticality level D overrun, the system drops tasks with criticality level E and use the second defined WCET for tasks with criticality level of D, C, B, and A. Note that tasks with criticality levels of C, B, and A have three, four, and five different WCETs, respectively, and we need to choose different n_i corresponding to each WCET for tasks. Upon each overrun, the system switches to the next criticality mode, drops tasks with LC levels, and sets the next WCET for the higher criticality tasks. It is important to mention that the above discussion is about a general MC system with multi-criticality levels. Therefore, an efficient decision highly depends on the applications, scheduling algorithms, and system requirements, and also the optimization problem would be more complex that must be solved. However, the basic idea and the proposed approach can still be applied. In addition to this design-time approach, a run-time approach can be applied to improve different WCETs of tasks in each mode based on the run-time behavior of tasks in each mode.

Nevertheless, although the QoS is improved by well adjusting the low WCET of HC tasks, the LC tasks are mostly dropped in the HI mode. In the next chapter, we will present an approach in order to improve the QoS of LC tasks more by analyzing the task dropping policy in the HI mode. We propose a heuristic where a QoS-aware parameter for each task is introduced, and we then provide a task-drop-aware scheduling analysis based on the new parameter.

References

1. S. Baruah et al. "The preemptive uniprocessor scheduling of mixed-criticality implicit-deadline sporadic task systems". In: *Proc. of Euromicro Conference on Real-Time Systems (ECRTS)*. 2012, pp. 145–154.
2. D. Liu et al. "Scheduling Analysis of Imprecise Mixed-Criticality Real-Time Tasks". In: *IEEE Transactions on Computers (TC)* 67.7 (2018), pp. 975–991.
3. H. Su, D. Zhu, and S. Brandt. "An Elastic Mixed-Criticality Task Model and Early-Release EDF Scheduling Algorithms". In: *ACM Trans. Des. Autom. Electron. Syst. (TODAES)* 22.2 (2016), pp. 1–25.
4. Charles Therrien and Murali Tummala. *Probability and random processes for electrical and computer engineers*. CRC press, pp. 190, 2018, p. 190.
5. Clément Ballabriga et al. "OTAWA: an open toolbox for adaptive WCET analysis". In: *IFIP International Workshop on Software Technolgies for Embedded and Ubiquitous Systems*. Springer. 2010, pp. 35–46.
6. Z. Guo et al. "Uniprocessor Mixed-Criticality Scheduling with Graceful Degradation by Completion Rate". In: *Proc. on IEEE Real-Time Systems Symposium (RTSS)*. 2018, pp. 373–383.
7. D. Liu et al. "EDF-VD Scheduling of Mixed-Criticality Systems with De-graded Quality Guarantees". In: *Proc. on IEEE Real-Time Systems Symposium (RTSS)*. 2016, pp. 35–46.
8. Chuancai Gu et al. "Partitioned mixed-criticality scheduling on multiprocessor platforms". In: *Proc. on Design, Automation & Test in Europe Conference & Exhibition (DATE)*. 2014, p. 292.
9. A. Hoseinghorban et al. "CHANCE: Capacitor Charging Management Scheme in Energy Harvesting Systems". In: *IEEE Transactions on Computer-Aided Design of Integrated Circuits and Systems (TCAD)* 40.3 (2021), pp. 419–429. DOI:10.1109/TCAD.2020.3003295.
10. Michael Mitzenmacher and Eli Upfal. *Probability and computing: Ran- domization and probabilistic techniques in algorithms and data analysis*. Cambridge university press, 2017.

11. Wassily Hoeffding. "Probability inequalities for sums of bounded random variables". In: *The collected works of Wassily Hoeffding*. Springer, 1994, pp. 409–426.
12. M. R. Guthaus et al. "MiBench: A free, commercially representative embedded benchmark suite". In: *Proc. IEEE International Workshop on Work-load Characterization. WWC-4*. 2001, pp. 3–14. DOI: 10.1109/WWC.2001.990739.
13. Frank J Massey Jr. "The Kolmogorov-Smirnov test for goodness of fit". In: *Journal of the American statistical Association* 46.253 (1951), pp. 68–78.
14. Eric B Holmgren. "The PP plot as a method for comparing treatment effects". In: *Journal of the American Statistical Association* 90.429 (1995), pp. 360–365.
15. Sanjoy Baruah et'al. "Scheduling real-time mixed-criticality jobs". In: *IEEE Transactions on Computers (TC)* 61.8 (2012), pp. 1140–1152.
16. Behnaz Ranjbar et al. "Power-Aware Runtime Scheduler for Mixed-Criticality Systems on Multicore Platform". In: *IEEE Transactions on Computer-Aided Design of Integrated Circuits and Systems (TCAD)* 40.10 (2021), pp. 2009–2023. DOI: 10.1109/TCAD.2020.3033374.
17. Siva Satyendra Sahoo, Bharadwaj Veeravalli, and Akash Kumar. "CL(R)Early: An Early-stage DSE Methodology for Cross-Layer Reliability-aware Heterogeneous Embedded Systems". In: *Proc. on ACM/IEEE Design Automation Conference (DAC)*. 2020, pp. 1–6. DOI: 10.1109/DAC18072.2020.9218747.
18. A. Yazdanbakhsh et al. "AxBench: A Multiplatform Benchmark Suite for Approximate Computing". In: *IEEE Design & Test* 34.2 (2017), pp. 60–68. DOI: 10.1109/MDAT.2016.2630270.
19. Zaid Al-bayati et al. "A four-mode model for efficient fault-tolerant mixed-criticality systems". In: *Proc. on Design, Automation & Test in Europe Conference & Exhibition (DATE)*. 2016, pp. 97–102.
20. G. Chen et al. "Utilization-Based Scheduling of Flexible Mixed-Criticality Real-Time Tasks". In: *IEEE Transactions on Computers (TC)* 67.4 (2018), pp. 543–558.
21. Z. Guo, L. Santinelli, and K. Yang. "EDF Schedulability Analysis on Mixed-Criticality Systems with Permitted Failure Probability". In: *Proc. of the International Conference on Embedded and Real-Time Computing Systems and Applications (RTCSA)*. 2015, pp. 187–196.
22. Félix-Antoine Fortin et al. "DEAP: Evolutionary Algorithms Made Easy". In: *J. Mach. Learn. Res.* 13.1 (July 2012), pp. 2171–2175.
23. B. Ranjbar et al. "Improving the Timing Behaviour of Mixed-Criticality Systems Using Chebyshev's Theorem". In: *Proc. on Design, Automation & Test in Europe Conference & Exhibition (DATE)*. 2021, pp. 264–269.
24. Behnaz Ranjbar et al. "BOT-MICS: Bounding Time Using Analytics in Mixed-Criticality Systems". In: *IEEE Transactions on Computer-Aided Design of Integrated Circuits and Systems (TCAD)* 41.10 (2022), pp. 3239–3251. DOI: 10.1109/TCAD.2021.3127867.
25. S. Pagani et al. "Machine Learning for Power, Energy, and Thermal Management on Multicore Processors: A Survey". In: *IEEE Transactions on Computer-Aided Design of Integrated Circuits and Systems (TCAD)* 39.1 (2020), pp. 101–116.
26. Petros S Bithas et al. "A survey on machine-learning techniques for UAV-based communications". In: *Sensors* 19.23 (2019), p. 5170.
27. Gavin A Rummery and Mahesan Niranjan. *On-line Q-learning using connectionist systems*. Vol. 37. University of Cambridge, Department of Engineering Cambridge, UK, 1994.
28. Somdip Dey et al. "User interaction aware reinforcement learning for power and thermal efficiency of CPU-GPU mobile MPSoCs". In: *Proc. on Design, Automation & Test in Europe Conference & Exhibition (DATE)*. IEEE. 2020, pp. 1728–1733.
29. D. Biswas et al. "Machine learning for run-time energy optimisation in many-core systems". In: *Proc. on Design, Automation & Test in Europe Conference Exhibition (DATE)*. 2017, pp. 1588–1592.

Chapter 4
Safety- and Task-Drop-Aware Mixed-Criticality Task Scheduling

In Mixed-Criticality (MC) systems, the frequent deadline misses or service degradation of some Low-Criticality (LC) tasks, such as mission-critical tasks, in the HIgh-criticality mode (HI mode) may have a negative impact on the other High-Criticality (HC) tasks and mission-critical tasks themselves, and consequently on the entire system, and may prevent the system from accomplishing its mission correctly. Therefore, in this chapter, we propose a novel scheme in order to reduce the number of deadline misses of LC tasks in the HI mode through task dropping analysis. Since safety-critical tasks are vital and their failure has a more devastating effect than mission-critical ones, we consider safety-critical tasks as HC tasks, while mission-critical and noncritical tasks as LC tasks.

We propose *FANTOM* (**FA**ult tolera**N**t **T**ask-dr**O**p aware scheduling For **MC** systems), a novel heuristic, which is based on a newly defined QoS-aware parameter and scheduling analysis of MC tasks with different criticality levels, by considering safety requirements. In *FANTOM*, the schedulability analysis is conducted in an offline manner in order to guarantee that all tasks with different criticality levels are executed properly before their deadlines in the presence of faults and based on the operational mode of MC systems. Thus, the main objective of *FANTOM* is to execute the majority of the LC tasks in the HI mode by considering a maximum allowable number of drops for each LC task. In addition, we guarantee the safety requirement of all MC tasks in both LO mode and HI mode. This is despite the fact that most of the related works cannot guarantee the safety requirement in the HI mode. Furthermore, the proposed method can schedule more task sets (i.e., it has a higher acceptance ratio) compared to similar works [1]. In summary, the main contributions of this work are:

- Introducing a new task parameter, which is utilized to drop LC tasks consciously in the HI mode (i.e., by introducing a maximum allowable number of drops for every LC task)

B. Ranjbar et al., *Quality-of-Service Aware Design and Management of Embedded Mixed-Criticality Systems*, https://doi.org/10.1007/978-3-031-38960-3_4

- A novel heuristic (*FANTOM*) based on the introduced parameter and the scheduling policy in the HI mode, in which an MC task schedulability analysis is developed by considering safety requirements and fault tolerance

To the best of our knowledge, *FANTOM* is the first study of its kind, which considers the scheduling analysis of MC tasks in order to prevent frequent drops of LC tasks in the HI mode by assigning a predefined threshold to them, while the safety requirements of tasks are guaranteed.

In the rest of this chapter, the problem statement and motivational example are presented in Sect. 4.1. In Sect. 4.2, we describe *FANTOM* in detail, while our experimental results have been described in Sect. 4.3. Finally, we conclude the chapter in Sect. 4.4.

4.1 Problem Objectives and Motivational Example

Four objectives have been set to be achieved in this chapter:

1. All of the MC tasks should be executed by their deadlines in the LO mode.
2. In cases that the system switches to the HI mode, all of the HC tasks should be finished before their specified deadline.
3. It should be guaranteed that the mission-critical tasks should not be frequently dropped in the HI mode.
4. The non-mission-critical tasks (also known as non-criticality tasks) will be dropped in the HI mode in order to ensure that the HC tasks and mission-critical tasks meet their deadlines.

According to these objectives, since frequent dropping or postponing their execution for a long time in the HI mode is not appropriate, we first introduce a new parameter, which limits the number of drops per LC task. We assign a new parameter δ for each task, which determines the minimum interval between two consecutive drops, that is set to ($\delta_i \times T_i$, where T_i is the period of task τ_i). Since dropping HC tasks is prohibited, we have set $\delta_{HC} = \infty$, which means that no dropping is allowed for HC tasks. In addition, some LC tasks are noncritical or non-mission-critical and dropped in the HI mode. Therefore, we define $\delta = 1$ for these tasks. Based on the application, the value of δ for each MC task is determined by designers. In the following, a motivational example will be discussed in which a task set containing all types of tasks (based on criticality levels, defined for the avionic industry in Table 2.1) has been considered.

We are going to give a motivational example based on Fig. 4.1 to clarify the problem and our solution for limiting the number of frequent drops per LC task. In this regard, assume that a single core executes five MC tasks ($\tau_1, \tau_2, \ldots, \tau_5$). The timing parameters of each task are shown in Table 4.1. As mentioned in Sect. 2.1.1.1, the deadlines of independent tasks are set to their periods. Each instance (job) of a task τ_i must be executed in the task time period, and as a result, the task generates a

Table 4.1 Example of MC task set

	ζ_i	$WCET_i^{LO}$	$WCET_i^{HI}$	T_i	$\hat{d}_i$	δ_i
τ_1	HC	1	5	12	6	∞
τ_2	HC	1	2	24	12	∞
τ_3	LC	1	1	4	–	3
τ_4	LC	1	1	3	–	4
τ_5	LC	1	1	6	–	1

$^*\tau_1, \tau_2 \in \{A\}$, $\tau_3, \tau_4 \in \{B, C\}$ and $\tau_5 \in \{D, E\}$

sequence of jobs during its execution. In Fig. 4.1, activated job sequences for each task are shown by an upward arrow. Furthermore, for showing the kth instance (job) of a task τ_i, we use notation $J(i, k)$. In this example, the first two tasks (τ_1, τ_2) are HC tasks (from level A in safety-criticality requirements of avionic industry), while others are LC tasks. LC tasks consist of both mission-critical tasks, which are from levels $\{B, C\}$, and non-mission-critical tasks from level $\{D, E\}$. In addition, we assume that mission-critical tasks should not be dropped frequently in the HI mode due to the occurrence of catastrophic consequences (which we discussed at the beginning of this section). Hence, in this example, we focus on our scheduling policy, and we can suppose that the measurement of the execution time will be accomplished after applying the fault-tolerance technique [1]. In addition, in this example, all of the tasks should comply with their specified time budget ($WCET_i^{LO}$ in the LO mode and $WCET_i^{HI}$ in the HI mode) to be executed correctly. Our task set is able to be scheduled under EDF-VD [2]. Also, the virtual deadlines ($\hat{d}_i$) of the HC tasks are computed and mentioned in Table 4.1 (virtual deadlines are less than the actual deadlines and provide a higher priority for HC tasks). For the simplicity of this example, we round the virtual deadlines into acceptable integers. Suppose that, when the first job of the τ_1 (which is denoted as J(1,1)) is being executed, due to the occurrence of a transient fault (which may be caused by high-energy neutron or alpha particle strikes in integrated circuits and therefore silently corrupt the data and lead to incorrect computation results), the system switches to the HI mode. As shown in Fig. 4.1, the system switches at timeslot 3 in this example.

The operation of the EDF-VD scheduling algorithm (by considering task killing) to the task set under [3] and [1] has been shown in Fig. 4.1a. Whenever the system switches to the HI mode, the active jobs of LC tasks will be dropped until the system would safely switch back to the LO mode (at time 9 in Fig. 4.1a when there is no active HC task). According to Fig. 4.1a, the active jobs of LC tasks, which are mission-critical tasks (τ_4) (J(4,2) and J(4,3)), are dropped twice, which is not tolerable for these tasks (according to Table 4.1, this task can be dropped once in each $\delta_4 \times T_4$ in the HI mode).

On the other hand, consider a situation that mission-critical tasks (τ_3, τ_4) would not be dropped in the HI mode. Actually, the execution of these tasks becomes as important as the execution of HC tasks. Therefore, τ_5 as an LC task is the only task, which is dropped in the HI mode. Similar to the previous example, EDF-VD algorithm is applied to the task set. Based on Fig. 4.1b, when the system switches to

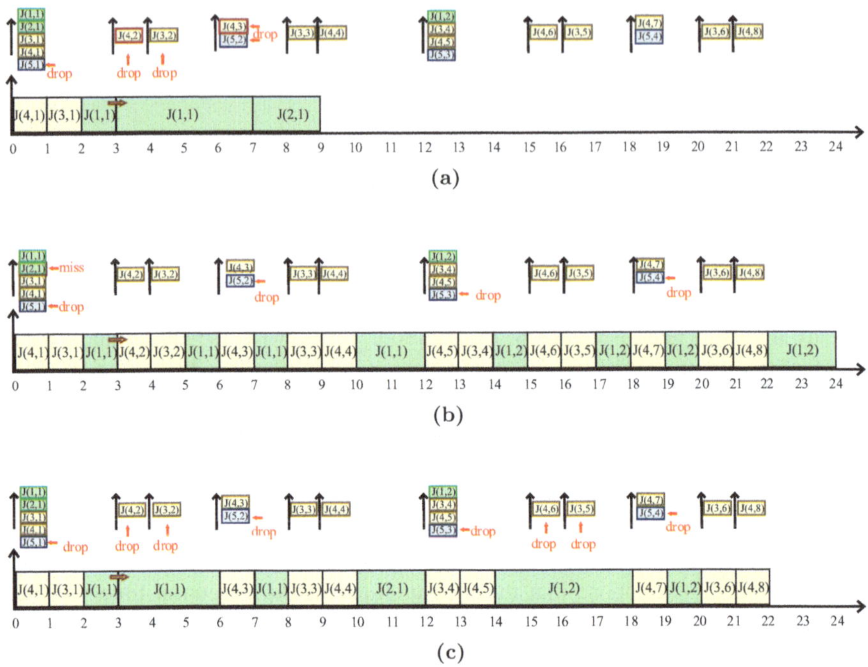

Fig. 4.1 Different schedules for the MC task set example within the interval [0, 24]. (**a**) The EDF-VD schedules where all LC tasks (mission-critical tasks and non-mission-critical tasks) are dropped in the HI mode according to [3] and [1]. (**b**) The EDF-VD schedules where mission-critical tasks (which are LC tasks) are not dropped in the HI mode. (**c**) The EDF-VD schedules under our policy

the HI mode, according to the EDF-VD policy (the ready tasks are sorted based on their deadline in ascending order to be executed on the core), the first job of HC task τ_2 (J(2,1)) is not executed due to lack of core space, in its predefined period and it will miss its deadline at time 24. Such scenarios of task scheduling may cause catastrophic consequences due to deadline missing of HC tasks.

Now, consider a parameter δ_i for LC tasks. When the system switches to the HI mode due to the execution time of HC tasks, LC tasks will be dropped in every δ_i to create slack time for the HC tasks to be executed before their deadlines. According to Fig. 4.1c, when the system switches to the HI mode, the first job of the mission-critical tasks τ_3 and τ_4 is dropped, which is J(3,2), J(4,2). Consequently, all of the HC tasks will be executed in this HI mode. Hence, since $\delta_5 = 1$, it would be always dropped in the HI mode. At the same time, by employing this technique, mission-critical tasks would not be frequently dropped, which is desirable.

4.2 FANTOM in Detail

In this section, first, we briefly introduce the quantification of MC tasks in Sect. 4.2.1. Then, in Sect. 4.2.2, we define MC task utilization, and based on them, we present the scheduling analysis technique and system upper bound utilization for the proposed heuristic (*FANTOM*) in Sects. 4.2.3 and 4.2.4, respectively. In the end, a general design-time scheduling algorithm is presented in Sect. 4.2.5, in which all essential conditions that must be guaranteed are determined.

4.2.1 Safety Quantification

As discussed in Sect. 2.1.2, in this book, the re-execution of the tasks has been used in order to tolerate transient faults. Independent from the level of criticality, any of the jobs in the tasks are executed up to n_i times to guarantee their safety requirements with regard to PFH, mentioned in Table 2.1. Any of the jobs in both LC and HC tasks requires maximum n_ζ times ($\zeta \in$ LC or HC) to be executed with regard to safety requirements [1]. Here, n_{HC} and n_{LC} are the maximum required re-execution times based on the PFH of both HC and LC tasks. However, in the worst case, if jobs execute n_i times to satisfy the safety requirements, it may cause the system to be overloaded (i.e., $U_{sys} > 1$) and lead the system to be unschedulable. To this end, in this chapter, analogous to the proposed mechanism in [4], we use a different policy based on the safety requirements to determine how the system mode switches happen and, therefore, how the low WCET of HC tasks are computed. Here, another time constraint n'_{HC} ($n'_{HC} < n_{HC}$) has been defined for HC tasks, which causes the system to operate without being overloaded. In addition, this time constraint gives the system the ability to switch to the HI mode when the correct response is not ready (e.g., due to a fault occurrence) after executing it for n'_{HC} times. In this case, we use our proposed drop-aware policy for LC tasks to guarantee the safe execution of HC tasks. Hence, n'_{HC} is the highest possible value that causes the system to be schedulable in the LO mode. Now, the WCET in each criticality level is computed as follows:

- LC tasks: $WCET_i^{LO} = WCET_i^{HI} = n_{LC} \times WCET_i$
- HC tasks: $\begin{cases} WCET_i^{HI} = n_{HC} \times WCET_i \\ WCET_i^{LO} = n'_{HC} \times WCET_i \end{cases}$

According to [1], the values of n_{HC} and n_{LC} for tasks in the same level of criticality are computed by solving Eq. (4.1), which has exploited the PFH of LC tasks and HC tasks based on Table 2.1, presented for the avionic industry. Hence, for each level, the value of PFH is the same from one hour to the next hour. Also, in Eq. (4.1), the HP is the hyper period of all tasks. Since the unit of the $pfh(\zeta)$ is hour, the HP′ represents the hyper period, in the unit of hour [1]. In this equation, $max(\lfloor \frac{HP - n_i \times WCET_i}{T_i} + 1 \rfloor, 0)$ represents the maximum number of execution rounds for task τ_i in the hyper-period $(0, HP]$. Moreover, f_i is a probability factor, which indicates the probability of an unsuccessful execution for a task due to transient faults (i.e., PoF). So, $f_i^{n_i}$ represents that a task is executed n_i times in the worst case to have successful execution but it fails in all executions. Therefore, the failure probability per hour for each criticality level ($pfh(\zeta)$) can be calculated by Eq. (4.1) [1]. As can be realized from this equation, by increasing the value of n_ζ (ζ= HC or LC), $pfh(\zeta)$ is decreased. Therefore, the minimum value of n_ζ for each criticality level is computed when $pfh(\zeta) \leq PFH_\zeta$:

$$pfh(\zeta) = \frac{(\sum_{\tau_i \in \tau_\zeta} max(\lfloor \frac{HP - n_\zeta \times WCET_i}{T_i} + 1 \rfloor, 0) \times f_i^{n_i})}{HP'} \tag{4.1}$$

In the next step, the value of n'_{HC} has to be computed in a way that the system would be schedulable and all the safety requirements are met. As we mentioned before, all LC tasks should be executed correctly before their deadline in the LO mode. Hence, by assigning n_{LC} to LC tasks, the number of re-executions for HC tasks in the LO mode (n'_{HC}) to have the schedulable system will be computed by solving Eq. (4.2) that $pfh(LO) < PFH_{LO}$ [1]:

$$pfh(LO) = \frac{(1 - \prod_{\tau_i \in \tau_{HC}} (1 - f_i^{n'_i})^{max(\lfloor \frac{HP - n'_i \times WCET_i}{T_i} + 1 \rfloor, 0)}) \times w(\infty, HP)}{HP'} \tag{4.2}$$

In this equation, the maximum number of execution rounds for HC task τ_i in each hyper-period ($[0, HP]$), that in each round, it is executed n'_{HC} times, is $max(\lfloor \frac{HP - n'_{HC} \times WCET_i}{T_i} + 1 \rfloor, 0)$. Since in the LO mode HC tasks are not executed more than n'_{HC} times to guarantee task schedulability, the probability that no job of HC tasks executes more than n'_{HC} times in each hyper-period is bounded by $P = \prod_{\tau_i \in \tau_{HC}} (1 - f_i^{n'_{HC}})^{max(\lfloor \frac{HP - n'_{HC} \times WCET_i}{T_i} + 1 \rfloor, 0)}$) that all HC tasks are executed successfully in maximum n'_{HC} times and no LC task is dropped. Therefore, the probability that LC tasks are dropped is $1 - P$. Due to the characteristics of our system, the maximum PFH for LC tasks, $w(\infty, HP)$, is defined and computed differently from what was defined in [1]. To compute this function, we should obtain the maximum number of executions for each task that can be done in one hyper-period. Normally, this number is accommodated by $max\{\left\lfloor \frac{HP - n_{LC} \times WCET_i}{T_i} + 1 \right\rfloor, 0\}$. Due to the newly defined parameter (δ_i) for the tasks, this round number is obtained and used in function $w(HP)$ as:

$$w(\delta, HP) = \sum_{\tau_i \in \tau_{LC}} max((\lfloor \frac{HP - n_{LC} \times WCET_i}{T_i} + 1 \rfloor - \lfloor \frac{HP - n_{LC} \times WCET_i}{T_i \times \delta_i} + 1 \rfloor), 0) \times f_i^{n_{LC}} \tag{4.3}$$

To find the maximum pfh for LC tasks in Eq. (4.2) that $pfh(LO) < PFH_{LO}$, parameter δ_i for each LC task in Eq. (4.3) should be infinitive. It means no LC task would be dropped in the LO mode. Therefore, as can be seen in Eq. (4.2), the value of n'_{HC} is independent of the values of δ_i of LC tasks that is used in Eq. (4.3). It should be noted that all HC tasks have the same value of n'_{HC} for their execution in the LO mode.

4.2.2 MC Task Utilization Bounds' Definition

In this section, we present the different utilization bounds for MC tasks, which are used in task scheduling. Based on the description of safety requirements presented in Sect. 4.2.1, the utilization of task j at level k is defined as $u_j^k = (WCET_j^k)/T_j$, in which if task j is an LC task, $WCET_j^k = n_{LC} \times WCET_j$ and if task j is an HC task, $WCET_j^k = n_{HC} \times WCET_j$ with k: LO, and $WCET_j^k = n'_{HC} \times WCET_j$ with k: HI. According to this definition, the low and high bound of utilization for different modes of task τ_j will be represented as u_j^{LO} and u_j^{HI}, respectively. Thus, the low-level and high-level utilization of HC tasks and also the low bound utilization of LC tasks are defined as follows:

$$\begin{cases} U_{HC}^{LO} = \sum_{\zeta_j=HCs} u_j^{LO} \\ U_{HC}^{HI} = \sum_{\zeta_j=HCs} u_j^{HI} \end{cases} \tag{4.4}$$

$$U_{LC}^{LO} = \sum_{\zeta_j=LC} u_j^{LO} \tag{4.5}$$

Theorem 4.1 *Due to the execution of some LC tasks, in the HI mode, the high bound utilization is presented as follows:*

$$U_{LC}^{HI} = \sum_{\zeta_j=LC} u_j^{LO} \times \frac{(\delta_j - 1)}{\delta_j} = u_j^{HI} \tag{4.6}$$

Proof Since we have to guarantee the correct execution of all HC tasks in the HI mode, a few LC tasks will be dropped in this mode. Therefore, we need to consider the jobs of LC tasks that are released in this HI mode. Due to the intended feature of LC tasks, one job could be dropped in every δ_j job instance in the HI mode. Accordingly, for these tasks, $\delta_j - 1$ jobs must be executed among δ_j jobs (i.e., if we have a period of time $\delta_j \times T_j$, LC tasks are executed for a time equal to $WCET_j^{HI} \times (\delta_j - 1)$ in this period, $u_j^{HI} = \frac{WCET_j^{HI} \times (\delta_j - 1)}{\delta_j \times T_j}$). Also, we have assumed that for each LC task, the value of $WCET_j^{LO}$ is equal to $WCET_j^{HI}$.

Therefore, the high bound utilization is rewritten as follows, which is lower than U_{LC}^{LO} (for each LC task, $u_j^{HI} = u_j^{LO} \times \frac{(\delta_j - 1)}{\delta_j} < u_j^{LO}$):

$$U_{LC}^{HI} = \sum_{\zeta_j = LC} (\frac{WCET_j^{HI} \times (\delta_j - 1)}{T_j \times \delta_j} = u_j^{LO} \times \frac{(\delta_j - 1)}{\delta_j} = u_j^{HI}) \tag{4.7}$$

Hence, in equality (4.6), if $\delta_j = 1$ for an LC task, then the utilization of this task j in the HI mode (u_j^{HI}) would be equal to 0. In other words, these tasks are not executed in the HI mode and then, *FANTOM* uses task dropping for these LC tasks with $\delta_j = 1$ (such as noncritical tasks) in the HI mode due to our proposed policy.

4.2.3 Scheduling Analysis

To guarantee the correct execution of MC tasks before their deadlines, several conditions have to be met at design time. We investigate the MC task schedulability in different system behavior as:

- Guaranteeing the task schedulability in the LO mode
- Guaranteeing the task schedulability in case of mode switching and then, in the HI mode
- Guaranteeing the task schedulability while the EDF-VD scheduling algorithm is used

The conditions for each item are explained in detail in this section. In the end, the last condition based on the system utilization is presented in Sect. 4.2.4. As mentioned before, at run-time, the MC system initially operates in the LO mode under EDF-VD. In *FANTOM*, when the system switches to the HI mode, with respect to the execution of HC tasks, the first job of LC tasks will be dropped. These LC tasks will be dropped periodically in an interval equal to $\delta_j \times T_j$ until the system is in the HI mode. In the meantime, all the HC tasks will be executed. Hence, if the mission of an LC task is more important than other LC tasks, the value of δ for this LC task would be higher. The details are as follows.

4.2.3.1 Conditions to Guarantee Task Schedulability in the LO Mode

By using the notations and definitions, we can easily express that MC task sets are schedulable under EDF if the following condition is guaranteed in a core in the LO mode:

$$U_{HC}^{LO} + U_{LC}^{LO} \leq 1 \tag{4.8}$$

As we have mentioned in Sect. 2.2.1, due to using the EDF-VD algorithm, deadlines of HC tasks will be downscaled by a multiplication factor x in the LO mode. Hence, the low bound utilization of HC tasks (U_{HC}^{LO}) would be downscaled by *1/x*. Based on the EDF-VD algorithm [3], due to the execution of LC tasks in the HI mode by using the parameter δ, we provide a problem formulation.

The following condition (which is obtained by the modification of the inequality (4.8)) is sufficient to schedule all of the tasks by the EDF-VD algorithm in the LO mode [2]. As mentioned, $U_{HC}^{LO} = \sum_{\zeta_i = HC} \frac{n_{HC} \times WCET_i}{T_i}$. Since $d_i = T_i$, and the virtual deadline $\hat{d_i} = x \times d_i$ is used for HC tasks in the LO mode to schedule tasks, therefore the utilization of HC tasks in the LO mode is $\frac{U_{HC}^{LO}}{x}$ that is used for task schedulability test:

$$\frac{U_{HC}^{LO}}{x} + U_{LC}^{LO} \leq 1 \tag{4.9}$$

4.2.3.2 Conditions to Guarantee Task Schedulability in the HI Mode

Since the MC systems must work successfully in the HI mode, we introduce a theorem and conditions to guarantee the deadline meeting of both HC tasks and LC tasks in the HI mode. Before introducing a new theorem, we explain how the new parameter δ is used. As depicted in Fig. 4.2, when the system switches to the HI mode, the demand requested for core computation time by the tasks is increased. In this figure, r_{ij} represents the release time of job j of task τ_i. In such critical situations (HI mode), the first job of the LC tasks is dropped, and consequently, one job of each τ_i is dropped every δ_i until the system switches back to the LO mode. These generated slacks (the difference between two r_is in the HI mode, where mission-critical tasks are not executed) are used to execute HC tasks in the HI mode. With this value of δ for each task, we determine a theorem for ensuring that both HC tasks and LC tasks (with $\delta > 1$) are scheduled within their deadlines, in the HI mode.

Theorem 4.2 *The sufficient establishing condition for executing both HC tasks and LC tasks in the HI mode is presented in inequality* (4.10), *in which, in the worst case, the system remains in the HI mode for the whole hyper-period. In this inequality, the HP is the hyper-period of HC tasks and LC tasks with* $\delta > 1$ *in a processing unit:*

$$\frac{\sum_{\zeta_j \in HC} \lfloor \frac{HP}{T_j} \rfloor \times WCET_j^{HI}}{HP} + \frac{\sum_{\zeta_j \in LC, \delta_j > 1} (\lfloor \frac{HP}{T_j} \rfloor - \lfloor \frac{HP}{T_j \times \delta_j} \rfloor) \times WCET_j^{HI}}{HP} \leq 1 \tag{4.10}$$

Proof Due to the execution of HC tasks and LC tasks in the HI mode, assume that the first task, which starts its execution, is an HC task, and this task causes the system to switch to the HI mode. In the worst case, the system remains in the HI mode

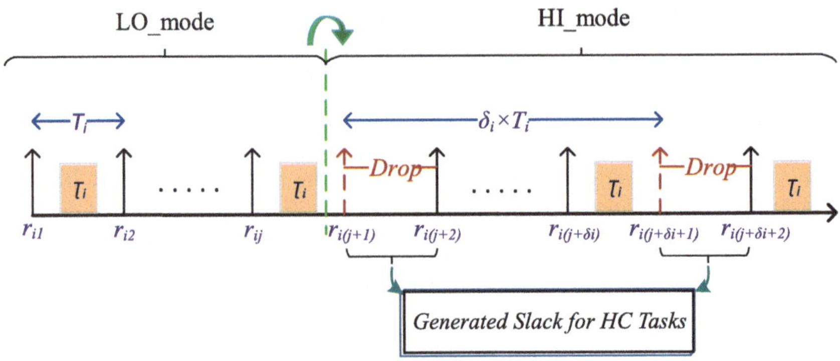

Fig. 4.2 LC task scheduling solution in both modes by dropping one job every δ_i in task τ_i in the HI mode

for the whole hyper-period. In this HI mode, the maximum time interval in which HC tasks are executed in one HP ($TInterval_{HC}^{max}$) is as:

$$TInterval_{HC}^{max} = \sum_{\zeta_j \in HC} \lfloor \frac{HP}{T_j} \rfloor \times WCET_j^{HI} \tag{4.11}$$

In addition, due to the execution of LC tasks in the HI mode and the nature of these tasks, this maximum time interval that LC tasks can be executed in one HP ($TInterval_{LC}^{max}$) is as:

$$TInterval_{LC}^{max} = \sum_{\zeta_j \in LC, \delta_j > 1} (\lfloor \frac{HP}{T_j} \rfloor - \lfloor \frac{HP}{T_j \times \delta_j} \rfloor) \times WCET_j^{HI} \tag{4.12}$$

Accordingly, if these two types of tasks need to be schedulable by the EDF algorithm in the HI mode before their deadlines, the following inequality must be guaranteed ($TInterval_{HC}^{max} + TInterval_{LC}^{max} \leq HP$):

$$\sum_{\zeta_j \in HC} \lfloor \frac{HP}{T_j} \rfloor \times WCET_j^{HI} + \sum_{\zeta_j \in LC \& \delta_j > 1} (\lfloor \frac{HP}{T_j} \rfloor - \lfloor \frac{HP}{T_j \times \delta_j} \rfloor) \times WCET_j^{HI} \leq HP \tag{4.13}$$

By dividing both sides of this inequality by HP, the inequality (4.10) will be obtained. Inequality (4.10) is the establishing condition for the schedulability of the tasks. ■

According to the EDF-VD algorithm, there are some scenarios in which the inequality (4.9) has been satisfied, while HC tasks in the HI mode have missed their deadline. Hence, a condition should be considered and satisfied in order to guarantee the schedulability of all of the HC tasks and LC tasks (based on the parameter δ) in their specified deadlines by the EDF-VD in the HI mode. Besides, there is a scenario that a task is released before mode switching, while its deadline is after mode switching and does not finish its execution yet, called *carry-over job* [5–7]. To consider the carry-over problem, the following sufficient condition has been expressed (the proof of this condition is the same lemma and has the same proving flow presented in [7, 8]):

$$U_{HC}^{HI} + (1 - x) \times U_{LC}^{HI} + x \times (U_{LC}^{LO}) \leq 1 \tag{4.14}$$

Proof To prove, suppose that τ_1 is an HC task with release time a_1 and deadline d_1 and causes the system switches to the HI mode at time t_1. Besides, τ_2 is an HC task that its deadline is missed at time t_2, while the system is in the HI mode ($0 < t_1 < t_2$). In addition, suppose that η_i is the cumulative execution time of each task τ_i in $[0, t_2]$. By this definition, we nominate $t_1 < (a_1 + x(t_2 - a_1))$. To prove this, consider that the absolute deadline of the HC task τ_1 is d_1 and its virtual deadline is $(a_1 + x(d_1 - a_1))$. As can be seen, τ_1 is overrun at time t_1 and continues to finish its execution completely before its deadline d_1, which is less than t_2 ($d_1 \leq t_2$). Thus, $t_1 < (a_1 + x(d_1 - a_1)) < (a_1 + x(t_2 - a_1))$

As we proposed, when the system switches to the HI mode, we drop the first job of each LC task τ_i, and then, they would be dropped every δ_i times as long as the system is in the HI mode. If an LC task with release time a_i and deadline d_i is released before time t_1, while its deadline d_i is after t_1 ($a_i < t_1 < d_i$), it is called carry-over job of LC tasks. For these carry-over jobs, we have $d_i < (a_1 + x(t_2 - a_1)$. To prove, it is obvious that the maximum cumulative execution time of an LC task τ_i is $(d_i - a_i)u_i^{LO}$, and it happens when the task can finish its execution before t_1. It means before overrunning of HC task τ_1. Since the EDF-VD algorithm is used for task scheduling, then $d_i \leq x \times d_1 \implies d_i \leq a_1 + x(d_1 - a_1)$. Since we have $d_1 < t_2$, therefore $d_i \leq a_1 + x(t_2 - a_1)$

Lemma *For any LC task τ_i:*

$$\eta_i \leq (a_1 + x(t_2 - a_1)) \times u_i^{LO} + (1 - x)(t_2 - a_1) \times u_i^{HI} \tag{4.15}$$

To prove, let us consider two cases: LC task τ_i is released within interval $(t_1, t_2]$ or it is not.

- *Case 1 (task τ_i is released within the time interval $(t_1, t_2]$):*
  - *With carry-over job ($t_1 < d_i$): The maximum cumulative execution time of LC task τ_i within the time interval $[0, t_2]$ is $\eta_i \leq (a_i - 0) \times u_i^{LO} + (d_i - a_i) \times u_i^{LO} + (t_2 - d_i) \times u_i^{HI} \implies \eta_i \leq d_i \times u_i^{LO} + (t_2 - d_i) \times u_i^{HI}$.*

    Since, we mentioned above, $d_i < (a_1 + x(t_2 - a_1))$, then $\eta_i \leq (a_1 + x(t_2 - a_1)) \times u_i^{LO} + (t_2 - (a_1 + x(t_2 - a_1))) \times u_i^{HI} \implies \eta_i \leq (a_1 + x(t_2 - a_1)) \times u_i^{LO} + (1 - x)(t_2 - a_1) \times u_i^{HI}$

- *No carry-over job* ($d_i < t_1$)*: The maximum cumulative execution time of LC task* τ_i *within the time interval* $[0, t_2]$ *is* $\eta_i \leq (t_1 - 0) \times u_i^{LO} + (t_2 - t_1) \times u_i^{HI}$.

  We mentioned that $t_1 < (a_1 + x(t_2 - a_1))$. *Therefore,* $\eta_i \leq (a_1 + x(t_2 - a_1)) \times u_i^{LO} + (t_2 - (a_1 + x(t_2 - a_1))) \times u_i^{HI} \implies \eta_i \leq (a_1 + x(t_2 - a_1)) \times u_i^{LO} + (1 - x)(t_2 - a_1) \times u_i^{HI}$

- *Case 2 (task* τ_i *is not released within the time interval* $(t_1, t_2]$*): It means the maximum cumulative execution of these tasks is* $\eta_i \leq (d_i - 0) \times u_i^{LO}$. *Now, we have two cases:* $d_i \leq t_1$ *and* $d_i > t_1$. *For the first one, since we mentioned,* $t_1 < (a_1 + x(t_2 - a_1))$, *and as we know that* $d_i \leq t_1$, *then* $\eta_i < t_1 \times u_i^{LO} \implies \eta_i \leq (a_1 + x(t_2 - a_1)) \times u_i^{LO} \implies \eta_i < (a_1 + x(t_2 - a_1)) \times u_i^{LO} + (1 - x)(t_2 - a_1) \times u_i^{HI}$. *Now, for the case of* $d_i > t_1$, *we proved that* $d_i < (a_1 + x(t_2 - a_1))$. *Thus,* $\eta_i < d_i \times u_i^{LO} \implies \eta_i \leq (a_1 + x(t_2 - a_1)) \times u_i^{LO} \implies \eta_i < (a_1 + x(t_2 - a_1)) \times u_i^{LO} + (1 - x)(t_2 - a_1) \times u_i^{HI}$

  Since we calculate the cumulative execution of tasks in the time interval $[0, t_2]$, *there are some LC tasks with* $\delta = 1$, *which are executed in the time interval of* $[0, t_1]$. *Therefore, the maximum cumulative execution of these tasks is* $\eta_i \leq (t_1 - 0) \times u_i^{LO}$. *Since* $t_1 < (a_1 + x(t_2 - a_1))$, *then* $\eta_i \leq (a_1 + x(t_2 - a_1)) \times u_i^{LO}$.

In addition, the maximum cumulative execution of HC tasks in the time interval $[0, t_2]$ can be computed as $\eta_i \leq \frac{a_1}{x} \times u_i^{LO} + (t_2 - a_1) \times u_i^{HI}$. It should be mentioned that HC tasks are executed with their virtual deadline in the time interval $[0, a_1]$, and since the HC task τ_1 overruns and the system switches to the HI mode, all tasks after a_1 will be executed by their actual deadline.

Now, let H denotes the cumulative execution of all tasks in the time interval $[0, t_2]$. Thus,

$$H \leq \sum_{\zeta_i \in LC \,\&\, \delta=1} (a_1 + x(t_2 - a_1)) \times u_i^{LO} + \sum_{\delta_i \in HC} \frac{a_1}{x} \times u_i^{LO} + (t_2 - a_1) \times u_i^{HI} +$$

$$\sum_{\zeta_i \in LC \,\&\, \delta>1} (a_1 + x(t_2 - a_1)) \times u_i^{LO} + (1 - x)(t_2 - a_1) \times u_i^{HI} \implies$$

$$H \leq (a_1 + x(t_2 - a_1)) \times U_{LC}^{LO}|^{\delta=1} + (a_1 + x(t_2 - a_1)) \times U_{LC}^{LO}|^{\delta>1} +$$

$$(1 - x)(t_2 - a_1) \times U_{LC}^{HI} + \frac{a_1}{x} \times U_{HC}^{LO} + (t_2 - a_1) \times U_{HC}^{HI} \implies$$

$$H \leq (a_1 + x(t_2 - a_1)) \times U_{LC}^{LO} + (1 - x)(t_2 - a_1) \times U_{LC}^{HI} + \frac{a_1}{x} \times U_{HC}^{LO} +$$

$$(t_2 - a_1) \times U_{HC}^{HI} \implies$$

$$H \leq a_1 \times (U_{LC}^{LO} + \frac{U_{HC}^{LO}}{x}) + x(t_2 - a_1) \times U_{LC}^{LO} + (1 - x)(t_2 - a_1) \times U_{LC}^{HI} + (t_2 - a_1) \times U_{HC}^{HI} \quad (4.16)$$

As presented in Eq. 4.9, $(U_{LC}^{LO} + \frac{U_{HC}^{LO}}{x}) \leq 1$, and then:

$$H \leq a_1 + x(t_2 - a_1) \times U_{LC}^{LO} + (1 - x)(t_2 - a_1) \times U_{LC}^{HI} + (t_2 - a_1) \times U_{HC}^{HI} \quad (4.17)$$

Hence, H is the maximum cumulative execution of all tasks. As we mentioned, τ_2 is one of these tasks that its deadline is missed. Therefore, H would be greater than t_2 (the time which τ_2 misses its deadline):

$$a_1 + x(t_2 - a_1) \times U_{LC}^{LO} + (1 - x)(t_2 - a_1) \times U_{LC}^{HI} + (t_2 - a_1) \times U_{HC}^{HI} > t_2 \implies$$

$$x(t_2 - a_1) \times U_{LC}^{LO} + (1 - x)(t_2 - a_1) \times U_{LC}^{HI} + (t_2 - a_1) \times U_{HC}^{HI} > t_2 - a_1 \implies$$

$$x \times U_{LC}^{LO} + (1 - x) \times U_{LC}^{HI} + U_{HC}^{HI} > 1 \quad (4.18)$$

Therefore, we must have the following inequality (4.14) to guarantee that no HC task misses its deadline in the HI mode:

$$x \times U_{LC}^{LO} + (1 - x) \times U_{LC}^{HI} + U_{HCT}^{HI} \leq 1 \quad \blacksquare$$

While the inequalities (4.14) and (4.10) are not necessary (just sufficient), the necessary condition would be driven when the sum of the utilization of LC tasks (i.e., mission-critical tasks) and HC tasks in the HI mode is higher than 1 or to guarantee the correct execution of jobs of each task before their individual deadlines in this HI mode. Thus, the necessary condition for scheduling both HC and LC tasks by the EDF algorithm in the HI mode and being executed correctly before their deadlines is the following condition:

$$U_{HC}^{HI} + U_{LC}^{HI} \leq 1 \quad (4.19)$$

4.2.3.3 Conditions to Guarantee Task Schedulability with EDF-VD Algorithm

Now, we present the value of x to obtain the virtual deadline by multiplying the actual deadline by x. Then, we present a new condition based on the previous conditions and the EDF-VD algorithm. By considering the inequalities (4.9) and (4.14), it could be concluded that the value of x ($d'_j = x \times d_j$) is obtained through the inequality (4.20):

$$\frac{U_{HC}^{LO}}{1-(U_{LC}^{LO})} \leq x \leq \frac{1-(U_{HC}^{HI}+U_{LC}^{HI})}{(U_{LC)}^{LO})-U_{LC}^{HI}} \tag{4.20}$$

Based on the interval for x in this inequality, and according to expression (4.21), the EDF-VD algorithm chooses the smallest value for x [2]. In addition, as explained before, we use the fault-tolerance technique, re-execution, to guarantee the correct execution of all tasks within their safety requirements based on Table 2.1 in any circumstance. To show how the parameters, such as safety requirements and virtual deadlines, affect each other, we can rephrase the value of x as follows. In this expression, when a task is executed and a fault occurs, it needs to be re-executed for a maximum of $(n_{\zeta_j} - 1)$ times to guarantee its safety requirement that ζ_j is LC or HC. Hence, the value of x is independent of δ_j. As mentioned in Sect. 4.2.1, the values of parameter δ_j has no effect on computing n'_{HC} (based on Eqs. (4.2) and (4.3)):

$$x \leftarrow \frac{U_{HC}^{LO}}{1-(U_{LC}^{LO})} \Longrightarrow x \leftarrow \frac{n'_{HC} \times \sum_{j\in HC} \frac{WCET_j}{T_j}}{1-(n_{LC} \times \sum_{j\in LC} \frac{WCET_j}{T_j})} \tag{4.21}$$

In addition to the mentioned conditions, another condition is required to guarantee that the task set would be schedulable by EDF-VD. In this regard, the upper bound utilization of the system will be computed by exploiting the condition (4.20) and represented as:

$$\frac{U_{HC}^{LO}}{1-U_{LC}^{LO}} \leq \frac{1-(U_{HC}^{HI}+U_{LC}^{HI})}{U_{LC}^{LO}-U_{LC}^{HI}} \Longleftrightarrow$$

$$U_{HC}^{LO} \times (U_{LC}^{LO} - U_{LC}^{HI}) \leq (1 - U_{HC}^{HI} - U_{LC}^{HI}) \times (1 - U_{LC}^{LO}) \Longleftrightarrow$$

$$U_{HC}^{HI} \leq 1 - U_{LC}^{HI} - \frac{U_{HC}^{LO} \times (U_{LC}^{LO} - U_{LC}^{HI})}{1-U_{LC}^{LO}} \tag{4.22}$$

According to inequality (4.22), the upper bound utilization of HC tasks in the HI mode (U_{HC}^{HI}) is obtained. If this condition is satisfied, the given task set is schedulable by the EDF-VD algorithm under the conditions in *FANTOM*.

Generally, inspired by the presented conditions, the conditions that should be investigated in a core to guarantee that a task set is schedulable under EDF-VD algorithm in *FANTOM* are inequalities (4.10) and (4.23). Condition (4.23) is obtained by using inequalities (4.8) and (4.22):

$$max(U_{HC}^{LO} + U_{LC}^{LO}, U_{HC}^{HI} + U_{LC}^{HI} + \frac{U_{HC}^{LO} \times (U_{LC}^{LO} - U_{LC}^{HI})}{1 - U_{LC}^{LO}}) \leq 1 \tag{4.23}$$

4.2.4 System Upper Bound Utilization

In this section, we present the system upper bound utilization in order to enable MC tasks to be schedulable by the *FANTOM*. In the end, we present the last condition that must be guaranteed. We nominate U_p as an upper bound for the task set, which should be schedulable in both HI mode and LO mode. This bound is defined as follows:

$$U_p = max(U_{HC}^{LO} + U_{LC}^{LO}, U_{HC}^{HI} + U_{LC}^{HI}) \tag{4.24}$$

We have explained that the condition (4.20) is sufficient for the task sets to be schedulable by the EDF-VD algorithm. Hence, by using conditions (4.20) and (4.24), the following expression could be derived. Here, our goal is to find the U_p, which still satisfies the following expression. Thereby, we have:

$$\frac{U_{HC}^{LO}}{1 - U_{LC}^{LO}} \leq \frac{1 - (U_{HC}^{HI} + U_{LC}^{HI})}{U_{LC}^{LO} - U_{LC}^{HI}} \tag{4.25}$$

Since we have $U_{HC}^{LO}+U_{LC}^{LO} \leq U_p \Rightarrow (U_{HC}^{LO} \leq U_p - U_{LC}^{LO}$ and also $U_{HC}^{HI}+U_{LC}^{HI} \leq U_p \Rightarrow 1 - (U_{HC}^{HI} + U_{LC}^{HI}) \leq 1 - U_p$:

$$\frac{U_p - (U_{LC}^{LO})}{1 - (U_{LC}^{LO})} \leq \frac{1 - U_p}{(U_{LC}^{LO}) - U_{LC}^{HI}} \tag{4.26}$$

This condition will be satisfied if and only if:

$$(U_{LC}^{LO})^2 - U_{LC}^{LO} \times (1 + U_{LC}^{HI}) + 1 + U_p \times (-1 + U_{LC}^{HI}) \geq 0 \tag{4.27}$$

Accordingly, if Eq. (4.28) is met, expression (4.29) will be obtained according to the expression (4.27) (which is always true for each of the low-criticality utilization in [0, 1)):

$$1 + U_p \times (-1 + U_{LC}^{HI}) = \frac{(1 + U_{LC}^{HI})^2}{4} \tag{4.28}$$

$$(U_{LC}^{LO} - \frac{1 + U_{LC}^{HI}}{2})^2 \geq 0 \tag{4.29}$$

By simplification of Eq. (4.28), it will turn into:

$$U_p = \frac{3 + U_{LC}^{HI}}{4} \tag{4.30}$$

Accordingly, it could be concluded that the upper bound (U_p) depends on the utilization of the LC tasks in the HI mode. It means (U_p) depends on the parameter of δ_j, which is different for each LC task ($U_{LC}^{HI} = \sum_{j \in LC} \frac{(\delta_j - 1) \times WCET_j}{(\delta_j \times T_j)}$). Since the U_p is the utilization bound of the system, which is run on a single-core processor, its maximum value is 1. In the case of $U_{LC}^{LO} + U_{HC}^{LO} < U_{LC}^{HI} + U_{HC}^{HI}$, then $U_p = U_{LC}^{HI} + U_{HC}^{HI}$. Therefore, according to equality (4.30), in addition to inequality (4.22), another condition and upper bound for U_{HC}^{HI} should be checked to guarantee the schedulability of a task set in the HI mode, if we have $U_{LC}^{LO} + U_{HC}^{LO} < U_{LC}^{HI} + U_{HC}^{HI}$, which is:

$$U_{HC}^{HI} \leq \frac{3(1 - U_{LC}^{HI})}{4} \tag{4.31}$$

4.2.5 A General Design Time Scheduling Algorithm

Now, we review the proposed approach algorithm at design-time and show which of the presented conditions need to be checked. The pseudo-code of our scheduling algorithm has been illustrated in Algorithm 4.1, which explains the mechanism of the schedulability test. In summary, at the beginning and according to Eq. (4.1), we calculate the re-execution profiles for each task (either high or low). Also, we calculate the minimum re-execution profiles for HC tasks through Eq. (4.2) (line 1). Subsequently, the utilizations are calculated (line 2). In addition, we calculate the maximum re-execution profiles for HC tasks through the schedulability test (line 3). If the maximum re-execution profile is more than the minimum one, we select this amount as n'_{HC} and consequently, the utilization of HC tasks in the LO mode is calculated. Otherwise, the algorithm will return a false value (lines 4–8). According to Algorithm 4.1, at the first stage, *FANTOM* evaluates the utilization bound in both LO mode and HI mode. If they are less than 1, it means the task set can be scheduled by the EDF in both modes (lines 10–13). Otherwise, the inequality (4.10) is evaluated in order to check whether both of the HC tasks and LC tasks with $\delta_j > 1$ (which are mission-critical tasks) are executed in the HI mode or not. In addition, the two mentioned conditions in inequality (4.23) and also Eq. (4.31) in case of $U_{LC}^{LO} + U_{HC}^{LO} < U_{LC}^{HI} + U_{HC}^{HI}$ will be evaluated to test the schedulability of the task set (line 15). If all the conditions are met, the virtual deadline coefficient

Algorithm 4.1 Design-Time Scheduling Method Pseudo Code

Schedulability Test(Task Set)

1: (n_{LC}, n_{HC}) are obtained by Eq. (4.1) & (n'_{HC}) by Eq. (4.2)
2: $(U_{HC}, U_{HC}^{HI}, U_{LC}^{LO}, U_{LC}^{HI}) \leftarrow$ *Util_Computation (taskset,n_{LC}, n_{HC})*
3: $n'_2 = \sup\{max(n \times U_{HC} + U_{LC}^{LO}, U_{HC}^{HI} + U_{LC}^{HI} + \frac{n\times U_{HI}\times(U_{LC}^{LO}-U_{LC}^{HI})}{1-U_{LC}^{LO}}) \leq 1\}$;
4: **if** $n'_{HC} < n'_2$ **then**
5:     $n'_{HC} = n'_2$
6: **else**
7:     **return** "The task set is not schedulable"
8: **end if**
9: $(U_{HC}^{LO}) \leftarrow$ *Util_Computation (taskset,n'_{HC})*
10: **if** $U_{HC}^{HI} + U_{LC}^{HI} \leq 1$ & $U_{HC}^{LO} + U_{LC}^{LO} \leq 1$ **then**
11:     $\hat{T}_i \leftarrow T_i$ for all tasks
12:     Schedule Task set with EDF Algorithm
13:     **return** "The task set is schedulable"
14: **else**
15:     **if** Eq. (4.10) & Eq. (4.23) are satisfied & $[(U_{LC}^{LO} + U_{HC}^{LO} < U_{LC}^{HI} + U_{HC}^{HI}$ & Eq. (4.31)) || $(U_{LC}^{LO} + U_{HC}^{LO} \geq U_{LC}^{HI} + U_{HC}^{HI})]$ **then**
16:         $x \leftarrow$ is computed by Eq. (4.21)
17:         **if** Eq. (4.14) is satisfied **then**
18:             $\hat{T}_i \leftarrow T_i \times x$ for each HC task
19:             Schedule Tasks with modified EDF-VD Algorithm
20:             **return** "The task set is schedulable"
21:         **else**
22:             **return** "The task set is not schedulable"
23:         **end if**
24:     **else**
25:         **return** "The task set is not schedulable"
26:     **end if**
27: **end if**

will be assigned with the minimum value (line 16). Hence, the virtual deadlines of HC tasks will be obtained by using the mentioned value in Eq. (4.21). Then, the sufficient condition of Eq. (4.14) is checked (line 17). If this equation is satisfied and the algorithm returns the true value, the task set will be scheduled by the EDF-VD algorithm (in which all LC tasks will be dropped in every δ_j in the HI mode (lines 18–20)). On the other hand, if none of the conditions are met, the task set cannot be scheduled and the algorithm will return a false value (lines 22 and 25).

4.3 Evaluation

In this section, the experimental results of the *FANTOM* are validated through extensive simulations on two case studies from the avionics domain presented in [1] and [9]. Then, we evaluate the impact of MC task's parameter variations in the schedulability test of the task sets.

4.3.1 *Evaluation with Real-Life Benchmarks*

4.3.1.1 First Case Study of Flight Management System (FMS)

Avionic real-life applications have been used in different papers to evaluate their presented methods [1, 10, 11]. To evaluate our method, we use the Flight Management System (FMS) application introduced in [1], which consists of seven tasks from level B and four tasks from level C of safety requirement table (Table 2.1). We consider the tasks from level B as HC tasks and the tasks from level C as mission-critical tasks. Therefore, there are no noncritical tasks in this task set. We can also define different values for the PoF of each task. To evaluate this part, this parameter is assumed to be 10^{-5} [1, 12]. In addition, the value of the skip parameter for all LC tasks is considered to be $\delta = 4$. According to Eqs. (4.1) and (4.2), the number of re-executions for all of the tasks is calculated and set to n_{LC}=n_{HC}=3 and n'_{HC}=2, respectively, to guarantee the safety requirements of the tasks without task killing and service degradation. Figure 4.3 represents the impact of *FANTOM* on the system schedulability. As shown in this figure, by increasing n'_{HC}, the utilization of the system will be increased due to the HC tasks' low utilization increment in inequality (4.23). In addition, the system will no longer be schedulable when $n'_{HC} > 2$. On the other hand, by increasing n'_{HC} (the redundancy for HC tasks needs to be increased), the PFH of LC tasks will be decreased and consequently, the system's safety could be improved. In essence, the probability of mode switching would be decreased and as a result, the LC tasks will be dropped less likely. Hence, by assigning $n'_{HC} = 2$, PFH of LC tasks will be 10^{-10} and the utilization will be 0.95, which is less than 1.

4.3.1.2 Second Case Study of Flight Management System (FMS)

Here, we investigate a set of real tasks, in which HC tasks have the safety requirement of level A. There is a case study for FMS application, introduced in [9], which consists of four tasks, three tasks from level A (responsible for executing the necessary control steps and essential for a reliable flight behavior) and one task from level B (responsible for detecting the objects). For this example, the number

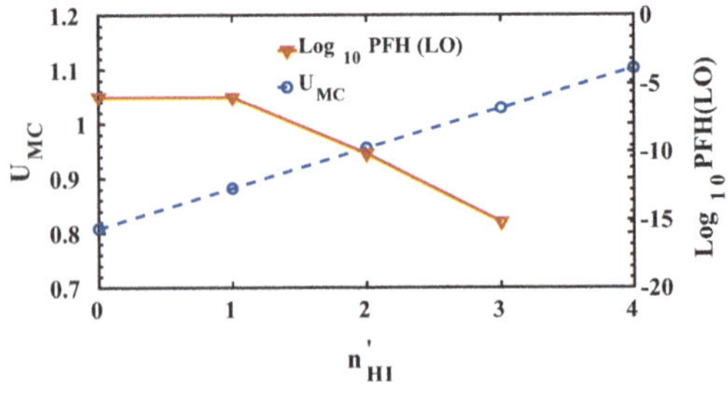

Fig. 4.3 FANTOM implementation for FMS application: case study 1

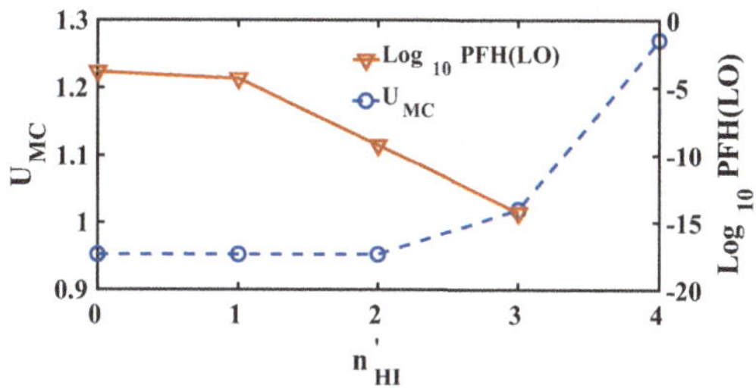

Fig. 4.4 FANTOM implementation for FMS application: case study 2

of re-executions is set to $n_{HC} = 3$, $n_{LC} = 1$, and $n'_{HC} = 2$, respectively. By having the same setting as the previous case study, Fig. 4.4 depicts the impact of *FANTOM* on system schedulability and LC task's PFH. As shown, the system is not schedulable for $n'_{HI} > 2$. Therefore, by assigning $n'_{HI} = 2$, PFH of LC task is 10^{-9} ($<10^{-7}$) and $U_{MC} = 0.96$.

In general, the safety requirement of HC (LC) tasks affects the number of re-executions ($n_{HC}(n_{LC})$), and consequently, the utilization and the number of re-executions for HC tasks in the LO mode (according to Eq. (4.2)) are changed. Hence, according to the Eq. (4.1), the number of re-executions for HC tasks and LC tasks depends on the task properties, such as WCET, PFH, and the number of tasks in each level. For example, for the same number of tasks and WCET for each task, if PFH is changed from 10^{-7} to 10^{-9} for the tasks' level, the number of re-execution may be increased to guarantee the safety requirement.

4.3.2 Evaluation with Synthetic Task Sets

4.3.2.1 Experimental Setup

In the following, the *FANTOM* has been evaluated by exploiting random MC task sets. These sets have been generated through the provided technique in [1, 2]. As an input parameter, the system's utilization (U_{bound}) is obtained as equality (4.24), which should be less than 1. In the beginning, the U_{bound} for the generated tasks are set to zero (i.e., the task system is initialized to be empty), and afterward, new tasks will be added to the task set in a random manner to increase the U_{bound} until a certain value (U_{bound} is increased with steps of 0.05). The period (T) and utilization of the tasks are generated uniformly within the range of [10, 100] and [0.01, 0.1], respectively. According to conditions, for each data-point (i.e., U_{bound}) in the range of [0.05, 1], 1000 task sets are generated and evaluated from the schedulability and fault-occurrence perspectives. In the end, the ratio of task sets, which were deemed as schedulable, will be reported.

In the established simulations, we have considered HC tasks from level A, LC tasks from level B to E, mission-critical tasks from level B or C, and non-mission-critical tasks from level D and E. Furthermore, the value of the parameter

(δ) is randomly generated between one and the maximum amount of this parameter for LC tasks. As the maximum amount of this parameter is considered to be determined by the designer, we investigate the results by varying the maximum amount in the range of [2,16] in the next subsection.

The efficiency of *FANTOM* has been investigated through extensive simulations and its comparison with the provided algorithms in [1] ([HYT14]) and [13] ([Al+16]). Researchers in [1] use EDF-VD to schedule the tasks. Also, researchers in [13] use the FP scheduling algorithm that is based on applying Response Time Analysis (RTA). In this regard, our observations are categorized into five subsections. There are some graphs relating to the results in which the y-axis represents the fraction of schedulable task sets, which is called the acceptance ratio, and the x-axis represents utilization. It should be noted that since the U_{bound} shows the utilization of task sets before applying the fault-tolerance technique and calculating the utilization in the LO mode and HI mode, the maximum utilization bound that the task set is schedulable has a small amount in graphs.

4.3.2.2 Effect of Varying LC Task's Parameter (δ)

In the beginning, we evaluate the effects of varying maximum value of the newly defined parameter (δ) for LC tasks. As a result, according to Fig. 4.5, by reducing the maximum value of parameter δ, the acceptance ratio will be slightly improved. The reason is that, in the case of increasing this parameter, the utilization of LC tasks in the HI mode will be increased due to Eq. (4.6). Therefore, the system schedulability will be decreased considerably according to inequality (4.23). If $max(\delta) = 1$, it means all LC tasks are noncritical and dropped in the HI mode. Indeed, there is no mission-critical task in the system. In this case, the acceptance ratio of the proposed approach is the same as the acceptance ratio of the method, proposed in [1], in which all LC tasks are dropped when the system switches to the HI mode. Indeed, there is no restriction for the system safety when none of the LC tasks are relevant to it. Besides, if $max(\delta) = 4$, it means LC tasks can have $\delta = 1$, $\delta = 2$, $\delta = 3$, or $\delta = 4$. Therefore, both noncritical tasks and mission-critical tasks are scheduled in the system as LC tasks. Hence, the significance of the proposed method is when there are some mission-critical tasks in the system, and in this case, the proposed method performs better than [1].

In the rest of this chapter, we consider $max(\delta) = 4$ to show the efficacy of our proposed method.

4.3.2.3 Acceptance Ratio of Schedulable Task Sets

We further compare the fraction of task sets, which could be scheduled under *FANTOM*, the traditional fault-tolerant EDF-VD algorithm [1], and fault-tolerant FP algorithm [13]. This fraction is defined as the acceptance ratio. Researchers in [1] have considered both task killing and service degradation for LC tasks in

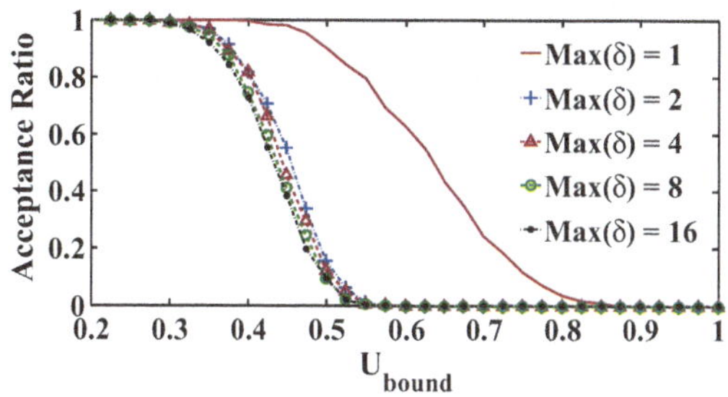

Fig. 4.5 Acceptance ratio with varying the parameter (δ) for LC tasks

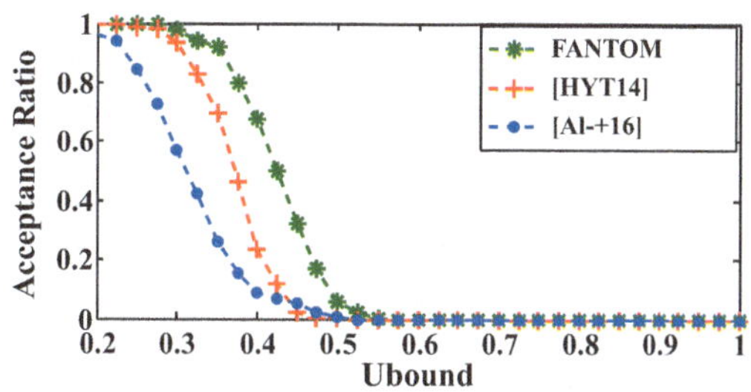

Fig. 4.6 Acceptance ratio of FANTOM with $max(\delta) = 4$ in comparison with methods of [1] (HYT14) and [13] (Al+16)

the HI mode. We use task killing of LC tasks by considering the QoS of LC tasks and execute them as much as possible in the HI mode to have a fair comparison. In addition, researchers in [13] have striven to increase the QoS of LC tasks in the HI mode, which we examine here. Accordingly, in this subsection, we have assumed that 40% of them are HC tasks and 60% are LC tasks that 30% of LC tasks are considered as mission-critical tasks and the remaining 30% are non-mission-critical tasks (with $\delta = 1$). It should be noted that a task set is schedulable if all tasks can be scheduled in the LO mode, and then after switching to the HI mode, all HC tasks would be executed and also, LC tasks cannot be frequently dropped. As shown in Fig. 4.6, when the utilization of the system is smaller than 0.225, the tasks are always schedulable by both algorithms, which use EDF-VD. By increasing the utilization bound, the proposed algorithm could always schedule the task sets as long as the utilization is smaller than 0.275. Furthermore, Fig. 4.6 explains that the proposed algorithm can improve the acceptance ratio by up to 43.9% and 65.9% compared to the traditional fault-tolerant EDF-VD algorithm [1] and fault-tolerant FP algorithm [13], respectively. Since the EDF-VD is used in both our proposed method and [1], in the rest of this chapter, we show the effectiveness of our proposed method in comparison with [1].

4.3.2.4 Effects of Using Fault-Tolerance Techniques

Now, we compare our proposed fault-tolerant scheme with the case when there is no fault-tolerant mechanism in the system. Indeed, we compare to a traditional non-MC scheduling algorithm in which the regular EDF algorithm is presented and

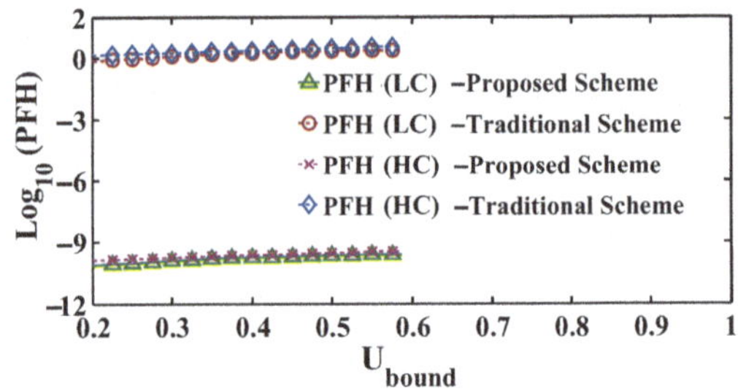

Fig. 4.7 Safety requirement guarantee for the system with and without fault consideration

applied in many previous studies [14]. Using the fault-tolerance techniques such as re-execution to guarantee the system's reliability and safety requirement has timing overheads, which is a common practice [15]. However, since the MC systems are safety-critical, its correct operation throughout a complete time interval is crucial even in the case of fault occurrence to prevent catastrophic consequences [16]. Figure 4.7 depicts that the proposed approach preserves the PFH of mission-critical tasks, and HC tasks, to less than 10^{-7} and 10^{-9} (introduced in Table 2.1), respectively, in any U_{bound} that the system is schedulable. In comparison, the traditional scheme has severely damaged the system's safety, which is not desirable. It should be noted that, as mentioned in the previous subsection, the system is not schedulable for $U_{bound} \geq 0.6$. Therefore, the results of the PFH for both mission-critical tasks and HC tasks are not shown for $U_{bound} \geq 0.6$.

4.3.2.5 Effects of HC Task Run-Time Behaviors ($P(WCET^{LO})$)

In this subsection, we evaluate the effect of changing the run-time behaviors of HC tasks on the acceptance ratio. $P(WCET^{LO})$ denotes the probability that HC tasks execute with their WCET in the LO mode ($WCET_i^{LO}$) (as discussed before, HC tasks may overrun and use their WCET in the HI mode). It can be seen that if the inequalities (4.10) and (4.23) are satisfied offline, the task set will be schedulable in both criticality modes. Hence, the schedulability of the task set is not affected by the variation of $P(WCET^{LO})$ in run-time.

4.3.2.6 Effect of Varying PoF for Task Instances

We evaluate the impact of varying PoF (f) on the system schedulability. Here, we assumed that 40% of tasks are HC tasks and 70% of the tasks are LC tasks (30% mission-critical task with $1 < \delta \leq 4$, and 30% non-mission-critical tasks with $\delta = 1$). As shown in Fig. 4.8, the acceptance ratio increases as (f) decreases from 10^{-5} to 10^{-7} and also from 10^{-7} to 10^{-9} in both our proposed method and the method proposed in [1]. The reason is that decreasing f means using a more reliable

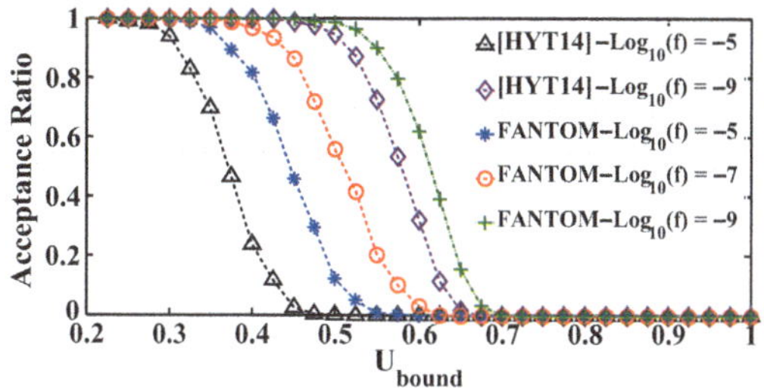

Fig. 4.8 Acceptance ratio with varying the PoF, and $max(\delta) = 4$ in FANTOM and method of [1] (HYT14)

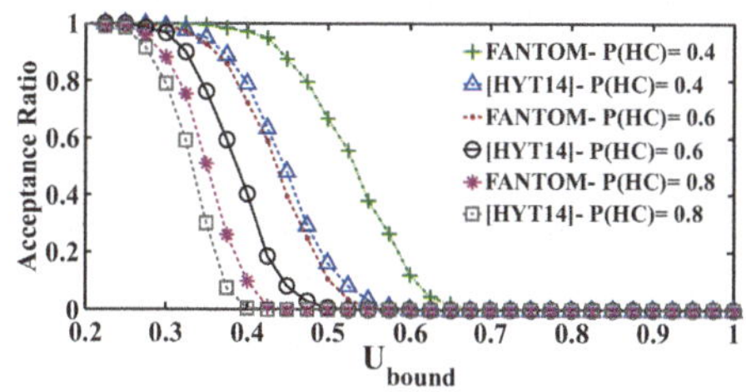

Fig. 4.9 Acceptance ratio with varying *P(HC)* in FANTOM and method of [1] (HYT14)

platform to have a safer system. However, the acceptance ratio of our proposed method is always better than the result of the proposed method in [1].

4.3.2.7 Effects of Task Mixtures with Varying P(HC) and P(MCT)

Now, we evaluate the effect of HC task distribution variation on the acceptance ratio. Note that since we are unaware of all real-life applications and the number of each type of critical level, we studied our proposed method's behavior with different parameter values. Here, $P(HC)$ denotes the ratio of HC tasks to all of the generated tasks. Here, in each scenario, we assume that the ratio of mission-critical tasks to all of the generated tasks is constant and the ratio of non-mission-critical tasks to all will be varied. Figure 4.9 shows that the acceptance ratio improvement becomes pronounced when there are fewer HC tasks in a task set (i.e., when P(HC) is decreased). However, when there are fewer HC tasks in a task set, there will be lesser number of system switches to the HI mode, and even by the occurrence of a mode switch, the system will switch back in earlier time (i.e., after a relatively shorter period of time). Consequently, the HI mode and LO mode will overlap less in time. In addition, LC tasks would fail less. This reasoning can also be obtained by exploring the condition (4.23). Besides, the same trend is found for [1], except that the proposed schemes always perform better than [1].

Similar to the above case, in Fig. 4.10, we evaluate how the number of mission-critical tasks affects the acceptance ratio (shown by $P(MCT)$). We assume that the ratio of HC tasks to LC tasks is constant and the ratio of the mission-critical tasks to non-mission-critical tasks in the task set is the only varying parameter. In addition, the distribution of HC tasks in a task set has been considered as a constant value (0.3). According to equality (4.30), since the upper bound utilization may

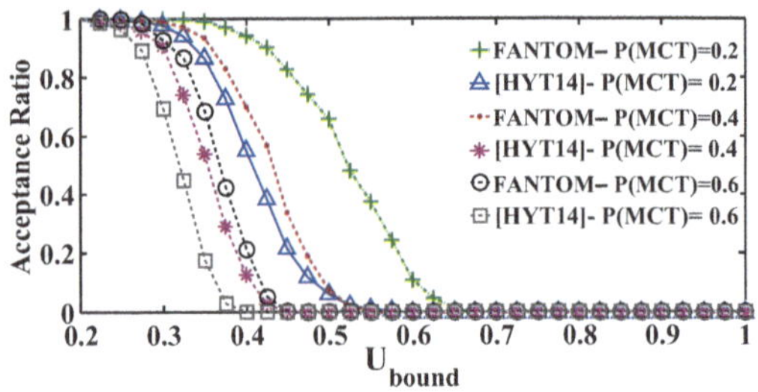

Fig. 4.10 Acceptance ratio with varying *P(MCT)* in FANTOM and method of [1] (HYT14)

be influenced by the mission-critical tasks, the acceptance ratio will be changed. According to Fig. 4.10, it is evident that we would have a noticeable amount of improvement as the utilization of mission-critical tasks is reduced. Decreasing the number of mission-critical tasks in a task set can cause a reduction in utilization and consequently, more tasks can be scheduled.

To this end, based on the results, we can conclude that by minimizing the ratio of HC tasks and mission-critical tasks in a task set, the upper bound will be maximized. In addition, the acceptance ratio of schedulable tasks would be increased by decreasing the task's PoF (i.e., a more reliable hardware platform is used).

4.4 Conclusions

This chapter presents a heuristic in which we introduced a new parameter, analyzed task-drop-aware scheduling for uni-processor MC systems, and guaranteed the safety requirements of MC tasks in the presence of faults. Existing tasks in these systems have different criticality levels from real-time and safety perspectives. In some MC systems, some LC tasks should not be frequently dropped in the HI mode to prevent catastrophic consequences. Therefore, by defining a new parameter (which specifies the minimum interval between two consecutive drops) that designers determine, we propose a task-drop-aware scheduling analysis based on the EDF-VD to schedule both types of tasks in the HI mode. We analyzed the results by varying different parameters in the system and obtained that the proposed method improves the acceptance ratio by up to 43.9% compared to the state of the art. Besides, in order to extend the proposed approach for five criticality levels (according to DO-178B standard), first, we have to know the importance of functions and how they can be dropped in the higher criticality modes to have no impact on system functionality. Then based on this knowledge, the MC task schedulability is analyzed and checked.

FANTOM guarantees the real-time constraints in the worst-case scenario. However, the system does not always exhibit the worst-case behavior at run-time. Besides, the proposed approach cannot adapt the system to task execution dynamism at run-time in order to optimize an objective. Therefore, we will propose a method in the next chapter to improve the QoS based on the dynamic changes in task execution times at run-time.

References

1. P. Huang, H. Yang, and L. Thiele. "On the scheduling of fault-tolerant mixed-criticality systems". In: *Proc. on Design Automation Conference (DAC)*. 2014, pp. 1–6.
2. S. Baruah et al. "The preemptive uniprocessor scheduling of mixed-criticality implicit-deadline sporadic task systems". In: *Proc. of Euromicro Conference on Real-Time Systems (ECRTS)*. 2012, pp. 145–154.
3. Sanjoy K Baruah et al. "Mixed-criticality scheduling of sporadic task systems". In: *European Symposium on Algorithms*. 2011, pp. 555–566.
4. Pengcheng Huang et al. "Energy efficient dvfs scheduling for mixed-criticality systems". In: *Proc. on Embedded Software (EMSOFT)*. 2014, pp. 1–10.
5. Sanjoy Baruah et al. "Scheduling real-time mixed-criticality jobs". In: *IEEE Transactions on Computers (TC)* 61.8 (2012), pp. 1140–1152.
6. Z. Guo et al. "Uniprocessor Mixed-Criticality Scheduling with Graceful Degradation by Completion Rate". In: *Proc. on IEEE Real-Time Systems Symposium (RTSS)*. 2018, pp. 373–383.
7. D. Liu et al. "Scheduling Analysis of Imprecise Mixed-Criticality Real-Time Tasks". In: *IEEE Transactions on Computers (TC)* 67.7 (2018), pp. 975–991.
8. D. Liu et al. "EDF-VD Scheduling of Mixed-Criticality Systems with De-graded Quality Guarantees". In: *Proc. on IEEE Real-Time Systems Symposium (RTSS)*. 2016, pp. 35–46.
9. P. Ittershagen, K. Gruttner, and W. Nebel. "Mixed-criticality system modelling with dynamic execution mode switching". In: *Proc. of Forum on Specification and Design Languages (FDL)*. 2015, pp. 1–6.
10. Michael Zimmer et al. "FlexPRET: A processor platform for mixed-criticality systems". In: *Proc. of Real-Time and Embedded Technology and Applications Symposium (RTAS)*. IEEE. 2014, pp. 101–110.
11. Steve Vestal. "Preemptive scheduling of multi-criticality systems with varying degrees of execution time assurance". In: *Proc. of Real-Time Systems Symposium (RTSS)*. IEEE. 2007, pp. 239–243.
12. Luyuan Zeng, Pengcheng Huang, and Lothar Thiele. "Towards the Design of Fault-tolerant Mixed-criticality Systems on Multicores". In: *Proc. of Compilers, Architectures and Synthesis for Embedded Systems (CASES)*. Pittsburgh, Pennsylvania, 2016, 6:1–6:10. ISBN: 978-1-4503-4482-1.
13. Zaid Al-bayati et al. "A four-mode model for efficient fault-tolerant mixed-criticality systems". In: *Proc. on Design, Automation & Test in Europe Conference & Exhibition (DATE)*. 2016, pp. 97–102.
14. Giorgio C Buttazzo. *Hard real-time computing systems: predictable scheduling algorithms and applications*. Vol. 24. Springer Science & Business Media, 2011.
15. Israel Koren and C Mani Krishna. *Fault-tolerant systems*. Morgan Kaufmann,2020.
16. A. Taherin, M. Salehi, and A. Ejlali. "Reliability-Aware Energy Management in Mixed-Criticality Systems". In: *IEEE Transactions on Sustainable Computing (TSUSC)* 3.3 (2018), pp. 195–208.

Chapter 5
Learning-Based Drop-Aware Mixed-Criticality Task Scheduling

As mentioned in the previous chapter, the low WCETs remain unchanged during run-time in static approaches, like *FANTOM*, which causes the system to be underutilized due to the unnecessary dropping of some LC tasks. Therefore, it is necessary to consider the run-time behavior of MC systems along with the assumptions that have been made at design-time (i.e., monitoring the state of the system and controlling the task dropping in the HI mode), to improve the utilization and QoS of LC tasks. Although there are some run-time approaches that improve the QoS by proposing a new scheduling policy or exploiting the dynamic slacks, the decision may be ineffective due to the lack of complete observation of the MC system's behavior, and there may be no guarantee of meeting the LC tasks' service requirements.

To this end, we propose a novel optimistic mechanism in this chapter that reduces the number of drops for the LC tasks by observing the system's behavioral changes at run-time. This goal has been achieved by exploiting the generated dynamic slacks in the decision-making process for the online task dropping to execute more LC tasks in the HI mode and enhance their schedulability. Since we are unaware of the amount of generated dynamic slacks during run-time in advance, ML approaches can be employed as a management technique for the prediction. Therefore, utilizing ML techniques as part of the proposed scheme has enabled it to partially exploit the dynamic slack to improve the QoS for the LC tasks in the HI mode. In these schemes, the learner finds the optimum drop rate for the LC tasks, prevents frequent drops in HI mode, and consequently reduces their deadline miss rate. We also extend the proposed mechanism, which is lenient in applying the learned drop-rate data to the scheduler. Accordingly, the main contributions of this chapter are:

- Presenting a novel adaptive technique with high QoS to schedule MC tasks at run-time
- Proposing a learning-based drop-aware MC task scheduling mechanism, called *SOLID*, to improve the QoS by exploiting the generated dynamic slacks rigorously, during run-time with no HC tasks' deadline misses

B. Ranjbar et al., *Quality-of-Service Aware Design and Management of Embedded Mixed-Criticality Systems*, https://doi.org/10.1007/978-3-031-38960-3_5

- Extending the proposed mechanism (*SOLID*) to a mechanism that uses accumulated dynamic slack moderately, called *LIQUID*

The rest of this chapter is organized as follows: In Sect. 5.1, we present a motivational example along with a problem statement to explain the problem to the readers better. The proposed approaches are discussed in detail in Sect. 5.2, and finally, we analyze and conclude the experiments in Sects. 5.3 and 5.4, respectively.

5.1 Motivational Example and Problem Statement

The main motivation for our proposed method comes from the fact that the MC systems are typically designed in a way that they are obliged to map and schedule the tasks in the worst-case scenario at design-time, before the system starts its operation. This is despite the fact that the application's QoS, system utilization, and deadline miss rate of LC tasks in case of HI mode switching will be affected while the application is executing at run-time. Indeed, the properties of MC systems can be improved according to the status of the tasks' execution over time. To support this claim, let us consider a simple drone application composed of five tasks ($\tau_1, \ldots, \tau_5$). The tasks' timing parameters have been shown in Table 5.1. The period of task τ_i (T_i) is equal to the task's relative deadline. Besides, since the EDF-VD algorithm is used to schedule the task, a virtual deadline ($\hat{d}_i$, which is less than the relative deadline) is needed to be defined for HC tasks (the detail of how it is computed has been explained in [1]).

In this example, τ_1, τ_2, and τ_5 are HC tasks and τ_3 and τ_4 are LC tasks. Each task function is determined in Table 5.1. Dropping an LC task, such as τ_3, could be acceptable in HI mode, but it should not frequently happen due to its responsibility. For instance, in multimedia tasks, e.g., τ_3, the maximum drop rate (i.e., the rate of skipping the videos) must be guaranteed when the MC systems are designed to satisfy the customers. More specifically, the minimum QoS of LC tasks must be guaranteed in the HI mode. Hence, most of the previously MC scheduling algorithms have been designed based on the maximum drop rate of LC tasks before the system starts its operation. Furthermore, these rates are kept constant during the tasks' operation at run-time when the system switches to the HI mode. Figure 5.1 illustrates four task scheduling approaches, including (1) method of [1], a design-time approach that drops all LC tasks in the HI mode; (2) *FANTOM* [2], the design-time approaches, which drop LC tasks based on their drop rates in the HI mode; (3) *FANTOM* [2], investigating their approaches at run-time; and finally, (4) proposed method in this chapter. According to this figure, each task has several jobs, released at the beginning of its period. Therefore, a released job (j) of task τ_i is shown by $J(i, j)$ with an upward arrow.

Figure 5.1a depicts the task scheduling procedure under the principles of the proposed mechanism in [1]. Accordingly, assume that all of the tasks should comply with their specified time budget ($WCET^{LO}$ and $WCET^{HI}$) to be executed

Table 5.1 The mixed-criticality task set

	Task function	ζ_i	$WCET_i^{LO}$	$WCET_i^{HI}$	T_i	$\hat{d}_i$	δ_i
τ_1	Engine control	*HC*	2	7	24	11	∞
τ_2	Collision avoidance	*HC*	2	4	48	22	∞
τ_3	Video capturing and transferring	*LC*	2	2	8	–	3
τ_4	Sensor data recording	*LC*	2	2	6	–	4
τ_5	Navigation	*HC*	0.8	1	12	6	∞

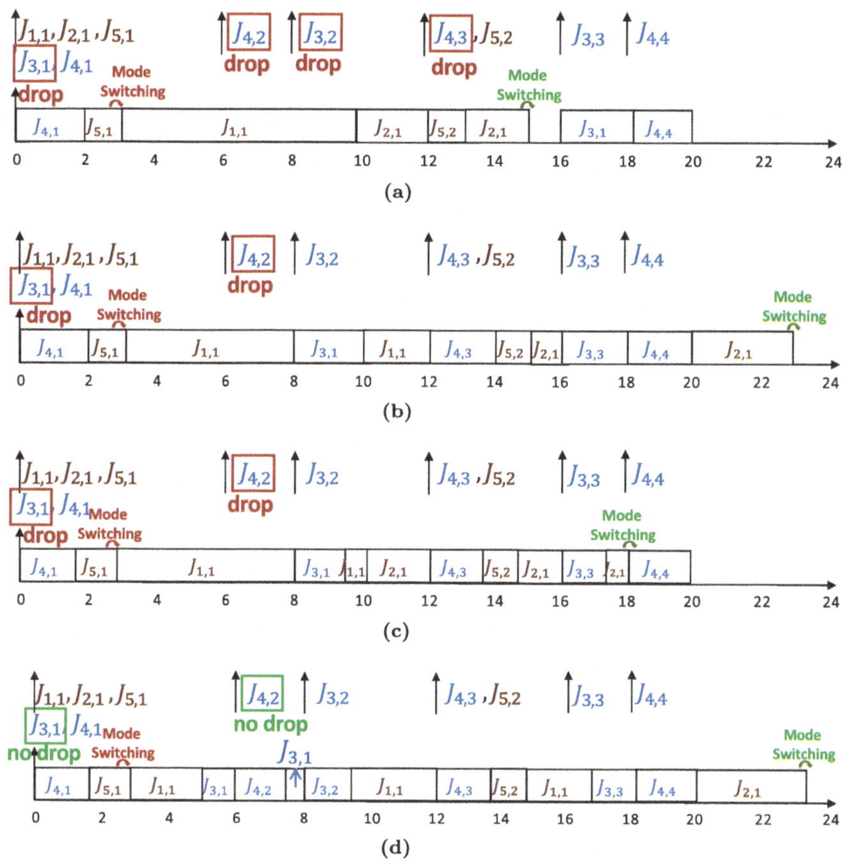

Fig. 5.1 Scheduling of MC tasks under different policies. (**a**) Task scheduling mechanism in [1] (worst-case scenario). (**b**) Task scheduling mechanism in [2] (worst-case scenario). (**c**) Task scheduling mechanism in [2] (run-time). (**d**) The proposed task scheduling mechanism in this chapter

correctly. This figure shows that the system switches to the HI mode by τ_5 overrunning. In this figure, the jobs of τ_3 ($J(3,1)$, $J(3,2)$), and τ_4 ($J(4,2)$, $J(4,3)$) are dropped twice in the HI mode, which is not acceptable in many MC applications. Figure 5.1b shows a scheduling mechanism that can schedule LC tasks in the

HI mode [2]. Accordingly, the LC tasks are dropped based on their predefined drop-rate parameters. Hence, as mentioned in Sect. 4.1, the drop-rate parameter is defined to limit the number of drops per LC task. In other words, it determines the minimum interval between two consecutive drops. In this scenario, only $J(3, 1)$, and $J(4, 2)$ will not be executed in the HI mode. As it can be seen, the presented approach in *FANTOM* [2] enables the MC system to schedule LC tasks in the HI mode and improves the LC tasks' QoS. Nevertheless, their design principles in the MC systems are all considered in the worst-case scenario of the task execution, which is not optimal. At run-time, the tasks are typically finished earlier than their WCET in most cases, and then some dynamic slack would be created. As an example, Fig. 5.1c shows the run-time behavior of the system, where some dynamic slack has been generated, and the tasks have finished their execution earlier. Therefore, other tasks could start their execution earlier, and the core would spend more time in the idle mode, compared to the scheduling mechanism depicted in Fig. 5.1b.

Based on what we have learned, it is recommended that the system should be able to manage its behavior during run-time, to minimize the drop rate of some LC tasks in the HI mode. As a result of this action, the QoS will be enhanced, e.g., less video will be skipped, which is desirable. Figure 5.1d represents the task scheduling at run-time, where the dynamic slack has been used to minimize the drop rate when the system is in the HI mode (such as $J(3, 1)$ and $J(4, 2)$ which are not dropped). Although we are not aware of the amount of dynamic slack in advance, they could be exploited to improve the QoS.

Motivated by the abovementioned example, prior to explaining the details of our novel scheduling technique, we define the constraints, and the objective function, as follows:

Deadline Constraints Each HC task τ_i with the WCET ($WCET_i^{LO/HI}$), running on core cr_j must finish its execution ($FTime_i$ is the finish time of task τ_i) correctly before its deadline (d_i) in both LO mode and HI mode. In addition, all LC tasks must finish their execution before their deadlines in LO mode. Besides, in case of switching to the HI mode, most LC tasks must finish their execution before their deadlines according to their drop rate δ_i:

$$
\begin{aligned}
&\forall \tau_i, \& \ \zeta_i = \text{HC} : FTime_i^{LO/HI} \leq d_i \\
&\forall \tau_i, \& \ \zeta_i = \text{LC} : \begin{cases} Mode = LO : FTime_i^{LO} \leq d_i \\ Mode = HI : FTime_i^{LO} \leq d_i | \delta_i \end{cases}
\end{aligned}
\tag{5.1}
$$

Objective Function We optimize the MC system QoS at run-time by maximizing the LC tasks' QoS in the system by utilizing the following objective function: *Maximize* QoS_{sys} or *Minimize* DMR_{sys}, where DMR is the deadline miss rate and the QoS is defined as the percentage of executed LC tasks in the HI mode to all LC tasks [2–4] ($QoS = n_L^{succ}/n_L$, where n_L is the number of all LC tasks and n_L^{succ} is the number of executed LC tasks before their deadlines in the HI mode), which can be optimized by optimizing their drop rates (δ_i) ($QoS_{sys} = n_L^{succ}/n_L$). The QoS is computed at the end of each hyper-period based on the number of non-executed LC tasks. Hence, the hyper-period is the LCM of all tasks' periods.

The problem is how to use the generated dynamic slack at run-time to maximize the QoS for satisfying the timing constraint. While we are not aware of the amount of dynamic slack in the future, it is possible to turn it into a partially controllable entity at run-time to optimize the intended objective. This could be done by using ML techniques. Therefore, this chapter proposes a novel learning-based and drop-aware scheduling for MC tasks running on a single-core platform. As discussed below, we utilize our newly proposed technique to use the generated dynamic slack at run-time to achieve better performance and QoS.

5.2 Proposed Method in Detail

In this section, we present SOLID, a novel **S**trict learning-**O**riented Qua**li**ty-of-Service- and **D**rop-aware task scheduling mechanism for MC systems, to apply in the run-time phase. As illustrated in Fig. 5.2, our proposed approach contains a design-time and a run-time phase. Application characteristics, architecture information, and QoS metric are counted as inputs and scheduled tasks, and QoS improvements are the outputs. Accordingly, we first need to analyze the task schedulability at design-time based on their characteristics (Sect. 5.2.1) to guarantee the minimum QoS requirement for the system. Therefore, we use the *FANTOM* approach to ensure the task schedulability and the minimum QoS. Finally, we exploit our newly introduced learning-based optimization mechanism at run-time (Sect. 5.2.2). In the following, we explain the details of the proposed approach in each of these phases.

5.2.1 An Overview of the Design-Time Approach

This section focuses on task scheduling at design-time. According to Fig. 5.2, the timing properties of tasks are obtained by running real-world benchmarks on a hardware platform (more details about the benchmarks and the platform are provided in Sect. 5.3). We validate the schedulability test using the tasks' parameters in the worst-case scenario. Besides, as part of ML technique employment, since embedded MC systems are the target systems, some aspects of the learning process, data, and model training are conducted at design-time, with robust offline learning techniques in the worst-case scenarios. More detail is discussed in Sect. 5.2.2.2.

In order to schedule the tasks on core, a well-known scheduling algorithm, EDF-VD, is used. Since this algorithm has been widely studied in previous studies, we mention this algorithm briefly in this and the next subsections. To test the task schedulability, the EDF-VD scheduling algorithm conditions are employed [1, 2].

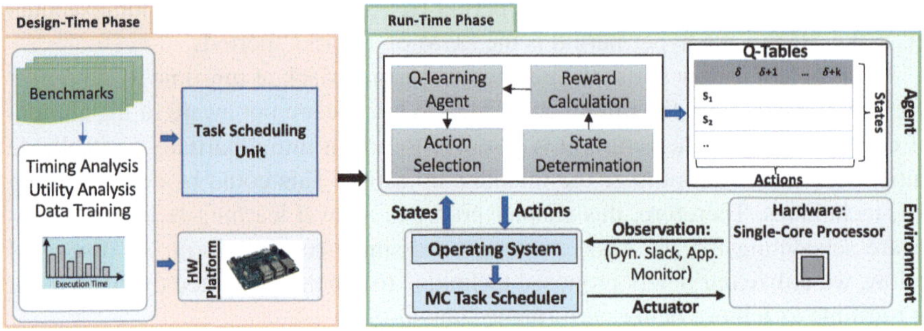

Fig. 5.2 An overview of design-time and run-time phases

Generally, the tasks are schedulable if all HC tasks can be executed correctly before their deadlines in any condition, and all of the LC tasks could be schedulable in the LO mode. In the HI mode, LC tasks can be scheduled based on their defined drop rates (i.e., the LC tasks' minimum QoS can be guaranteed in the worst-case scenario). As a result, a set of tasks is schedulable under the EDF-VD in both operational modes on each core if the following necessary and sufficient conditions are met [2]. Equation (5.2) presents the maximum utilization bounds for MC systems in both LO mode and HI mode that it must be less than one to let the tasks be schedulable under EDF-VD at run-time and also the system switch safely between the modes. Since the LC tasks must be dropped in the HI mode based on their drop-rate values with no effect on HC tasks' execution, Eq. (5.3) presents the sufficient condition for executing both HC and LC tasks in this HI mode (the information and the proofs of these conditions have been explained in detail in the previous chapter):

$$U^{MC} = max(U_{HC}^{LO} + U_{LC}^{LO}, U_{HC}^{HI} + U_{LC}^{HI} + \frac{U_{HC}^{LO} \times (U_{LC}^{LO} - U_{LC}^{HI})}{1 - U_{LC}^{LO}}) \leq 1 \tag{5.2}$$

$$\frac{\sum_{\zeta_j \in HC} \lfloor \frac{HP}{T_j} \rfloor \times WCET_j^{HI}}{HP} + \frac{\sum_{\zeta_j \in LC} (\lfloor \frac{HP}{T_j} \rfloor - \lfloor \frac{HP}{T_j \times \delta_j} \rfloor) \times WCET_j^{HI}}{HP} \leq 1 \tag{5.3}$$

where U_l^k denotes the total utilization of the tasks with the same criticality level l, in the mode k, and HP is the hyper-period of the tasks.

5.2.2 *Run-Time SOLID Approach*

The main goal of SOLID is to enhance the LC tasks' QoS (i.e., minimizing the number of dropped LC tasks) under the mode switching situation by exploiting the dynamic slacks at run-time, with no deadline misses of HC tasks in any situation. This capability has been brought to SOLID by exploiting learning-based techniques. Note that the learning algorithm and the scheduling algorithm are independent, and we do not use learning techniques to schedule the tasks. In fact, the proposed approach is independent of the task scheduling algorithm, and the learning process is used to improve the QoS independent of scheduling tasks. Here, any scheduling algorithm can be applied to the tasks; however, this learning-based task scheduling mechanism is built upon the EDF-VD algorithm. To guarantee the schedulability of HC tasks in both LO mode and HI mode, when the system begins its operation, HC tasks are scheduled based on their virtual deadlines, and the LC tasks are scheduled based on their actual deadlines. In the HI mode, all HC tasks are scheduled based on their actual deadlines, while the LC tasks will be scheduled according to the SOLID scheduling principles. It should be mentioned that although the learning process can be done independently of the scheduler, its timing overhead to obtain the new drop rate values of LC tasks may be significant in real-time systems. Therefore, we can consider its timing overhead while checking the task schedulability. The details of the timing overhead of the learning process and how it is considered in schedulability test are presented in Sect. 5.3.3. In this section, we fully describe SOLID, based on the mentioned reinforcement learning technique in Chap. 3.

5.2.2.1 Learning-Based System Property Optimization

As mentioned in Chap. 3, the general Q-learning/SARSA technique consists of the three main components [5, 6], including 1) a discrete set of states $S = \{s_1, s_2, \ldots, s_l\}$, 2) a discrete set of actions $A = \{a_1, a_2, \ldots, a_k\}$, and 3) reward function R. The algorithm collects the current state s_t and determines the next action a_t ($a_t \in A$). The Q-values are updated according to Eq. (3.21), based on the corresponding computed reward in every iteration. The algorithm learns the optimal action in every state, and this process is repeated until a predefined convergence criterion is met. Note that SARSA and Q-learning are two RL methods and tend to optimize the results in the end. However, SARSA is an online policy, while Q-learning is an offline policy. These two different policies lead to different next action selection and Q-table updating. Since the proposed approach is composed of offline and online phases, and it is critically important to have a robust offline training technique for considering the worst-case scenarios before the system gets operational at run-time, SARSA would be the best choice for finding the optimum value (because there is no urgency at design-time to find the optimum value). Therefore, SARSA can explore most of the states in the data training. In this work, we set the values of γ in Eq. (3.21) to 0.2, and α to 0.5. These values are determined based on a wide range of experiments, which are set to obtain the best improvement.

5.2.2.2 SOLID Optimization in Detail

In order to maximize the QoS in MC systems, here we propose our learning-based approach for the operation of the system in HI mode. It should be mentioned that the time of the system mode changes and how long the system stays in each mode are unknown. As a result, the system encounters dynamic slacks with varying lengths, which would result in different actions in the intended mode. In SOLID, the agent controller has been designed to maximize the QoS by decreasing the LC tasks' drop rates (i.e., drop less often) when the system switches to the HI mode. Note that no action is required in the learning process if the dynamic slack is too small. In order to update the Q-table, we check the generated dynamic slack at the end of each hyper-period. Based on the available slack, we update the Q-table, and decisions are taken (i.e., the new drop-rate value is determined). It should be mentioned that since we target real-time embedded MC systems, we conduct some parts of the learning process, data, and model training for the Q-table at design-time with robust offline learning techniques in the worst-case scenarios. The reason is that the learned Q-table can be utilized to quickly determine the optimal actions based on the system state and also reduce the probability of bad decisions. Then, by using this data and what the learning algorithm learns from this training phase at design-time and also the obtained historical data at run-time, the algorithm improves its prediction process as time passes. In the following, we first explain the system state, action determination, and reward computation for the learning process to generate the Q-table values. Then, we present the proposed approach in detail.

System State Determination There are various criteria for determining the system states. In our proposed scheme, for the Q-table, the states of the system depend on the available CPU utilization and dropped LC task in a period. To represent the states in a formal way, for each state s_i, $s_i = \|U_i^{MC}\| + \|\Delta_i\|$, where $U_i^{MC} = \{0, 0.1, \dots, 1\}$. Both utilization and dropped tasks are normalized to their maximum value (shown by $\|\dots\|$). We also define ten ranges to determine the percentage of missed tasks to all tasks. As an example, consider s_i as the ith utilization range ($max(i) = 10$). Therefore, we have $s_i + \Delta$ ($\Delta \in [1, 10]$), which indicates the variation in the rate of dropped LC tasks for the fixed utilization range. In each iteration, according to the unused utilization and the rate of dropped tasks in the previous hyper-period, the current state is determined for the Q-table. Here, we select an optimal action for the current system state. Thus, the system can gradually reach the optimal state.

Learning Action Determination In this method, the well-known ϵ-greedy policy has been exploited, in which the dynamic policy is used for adjusting ϵ [7]. We first use a dynamic ϵ-greedy policy with the value of 0.5 at design-time to prevent the probability of the learning algorithm from being stuck at few Q-values. Accordingly, we can accelerate the learning process. Afterward, the fixed ϵ-greedy policy is used with the value of 0.2 at run-time to ensure that the system reaches the optimum state and chooses the best action based on the Q-values, which has the maximum value. The action space in the Q-table illustrates an increase/decrease in a LC task's drop rate (δ_i,.., $\delta_i + k$). It should be noted that the minimum values of drop rates are equal to the initial values that is used for schedulability analysis at design-time.

Reward Computation This approach calculates the reward at the end of each hyper-period based on the available dynamic slack. Hence, when there is less accumulated dynamic slack at the end of the hyper-period, it means more core capacity ($U^{MC}(t)$) has been used on that hyper-period. The considered reward function for the Q-table is shown in Eq. (5.4), which is based on the generated dynamic slack at the end of each period:

$$R = \begin{cases} -\Gamma & U^{MC}(t) > \varphi \\ \frac{1}{10\times(1-U^{MC}(t))} & U^{MC}(t) < \varphi \\ +\Gamma & U^{MC}(t) = \varphi \end{cases} \tag{5.4}$$

The reward function considers three scenarios. If the utilization falls into the unsafe zone that may cause deadline violation, the decision will be penalized. An unsafe zone means the utilization may increase more than one, where EDF-VD cannot guarantee the timeliness of all tasks. Accordingly, it results in a negative value ($-\Gamma$, where $\Gamma > 0$ and has a constant value) for the reward function, which decreases the Q-value in Eq. (3.21), i.e., reduces the probability of choosing it in the future. In Eq. (5.4), we set the value of Γ equal to 100 to highly impact the value of the reward function in a negative manner. In addition, since our goal is to use all of the accumulated dynamic slack to optimize the system property, $U_t^{MC} = 1$ would be the optimum case for the reward function (presented in Eq. (5.2)). However, since there may be some errors in the first phases of the learning technique, we consider the upper bound of core utilization (φ) to be less than one ($\varphi < 1$). In fact, it can be equal to $\varphi = 1 - \mu$, where μ has an extremely low positive value, such as 0.05. Hence, we consider a value less than the maximum core utilization as the core utilization limit to never let it violate the threshold.

Actual Execution Time Predictor Policy Due to releasing several jobs of a task in each hyper-period, the execution times of jobs may be different in each hyper-period. We have to predict the execution times to compute the core utilization ($U^{MC}(t)$) in Eq. (5.2), according to the previous run-time tasks' execution times. This prediction is based on Eq. (3.23), mentioned in Chap. 3.

Figure 5.2 depicts our proposed learning-based drop-aware task scheduling mechanism. It consists of the environment, i.e., the hardware platform, the agent, and its interaction with the operating system and the applications. This learning-based property improvement technique has been designed for a system based on its states and action determination algorithms discussed earlier. The scheduler schedules the tasks based on the EDF-VD at run-time. At the end of each hyper-period, the accumulated dynamic slack and the number of dropped LC tasks in the HI mode are observed. The proposed learning phase decides how to increase/decrease the LC tasks' drop rates based on the Q-table and reward function value. The major goal of SOLID is to use most of the created slack time and consequently maximize the core utilization by optimizing the LC tasks' drop rates. In the learning process, the agent observes the state at a time period instance T_t, computes the award, updates the Q-table, and performs an action. The action (a) is selected from the predefined action set in the specified Q-table ($a \in \{a_1, a_2, \ldots, a_k\}$, where k is the maximum number of actions corresponding to each table). The chosen action is applied for the next time period (T_{t+1}). After decoding the actions for the HI mode by the operating systems, based on the new LC task drop rates, the policies of the task scheduling and LC tasks dropping will be updated in the case that the system mode switches to the HI mode.

To guarantee meeting the deadline of HC tasks, although we try our best effort and define the hard margin to avoid missing the deadlines, it could happen for the Earliest Deadline First (EDF) algorithm in the worst-case scenario that the core is fully utilized and all tasks are executed up to their WCETs at run-time, while some LC tasks' drop rates were increased according to learned data. It may lead to some deadline misses for HC tasks. As a result, SOLID is strict in applying the learned data into the scheduler to ensure meeting the deadlines in the worst-case scenario. The scheduler in SOLID approach always considers the initial drop-rate values (δ_i^{old}) and drops LC tasks based on them in the HI mode. To apply the learned data, the dynamic slack is detected at run-time. When an HC task finishes its execution early, a dynamic slack is generated due to the early completion. Based on the learned drop rate values, the scheduler in SOLID releases the LC jobs to execute in this generated dynamic slack. Therefore, LC tasks are executed more times (drop fewer) by exploiting the slack time generated only from the early completion of HC tasks' executions and improving the QoS in the HI mode. Although it introduces an extra workload for the system, it causes to prevent affecting the early LC tasks' releases on HC tasks' timeliness. We require judicious slack management to determine whether it is feasible to release an LC job at a time point.

However, in order to be lenient in applying the learned drop-rate data into the scheduler, we extend SOLID to LIQUID, which uses accumulated dynamic slack moderately to improve QoS.

5.2.3 Run-Time LIQUID Approach

In LIQUID (**L**earn**i**ng-based **Qu**al**i**ty-of-service- and **D**rop-aware MC task scheduling mechanism), like what we proposed in SOLID, the proposed learning algorithm operates independently of the scheduler. However, in contrast to SOLID, LIQUID applies the learned data into the scheduler with no restriction and taking care of generated dynamic slack by HC tasks. The scheduler in LIQUID approach always considers the learned drop-rate values, and LC tasks are dropped based on them in the case of mode switches.

To make both approaches explicit, consider a task τ_i with $\delta_i^{old} = 3$, which means one job would be dropped among three jobs in the HI mode. If the learned drop rate after a hyper-period is equal to $\delta_i^{new} = 6$, it means one job would be dropped among six jobs. Therefore, the rate of task dropping would be half in comparison with the initial value. In fact, one more job among six jobs would be scheduled and executed. LIQUID always considers δ_i^{new} for the LC task τ_i when the system switches to the HI mode. However, in SOLID, δ_i^{old} considers for task τ_i. If there is sufficient accumulated dynamic slack at run-time to execute τ_i based on the learned δ_i^{new}, the task is released. In this work, we exploit and adapt the early release policy, presented in [8], which is an effective slack management technique in MC systems and based on a known mechanism, called wrapper-task mechanism [9, 10].

Algorithm Algorithm 5.1 illustrates the pseudo-code of the run-time approach, including both scheduling and learning procedures at the same time. As inputs, the algorithm takes the tasks and their characteristics (e.g., WCET, criticality level, drop rate, and period), the hardware platform, and the minimum QoS requested by the tasks. In addition, since a part of the learning process is done at design-time, the Q-table is obtained and taken as input. On the other hand, improvements in the LC tasks' QoS and the scheduled tasks are defined as outputs at the end (*Time*). At each time, the scheduler checks the status of the tasks, whether they are overrun or not, which results in mode switching (line 4). This unit also checks the periods of tasks, whether they would be released or not. All tasks on every core are scheduled based on the EDF-VD algorithm (line 5). In the case of mode switches to the HI, the LC tasks are dropped based on their defined drop rate values to guarantee the correct execution of HC tasks. In lines 7–12, the number of dropped LC tasks is counted to be used in the learning process. If the system switches back to the LO mode, a parameter (*CountDrop*) for each task, which counts the number of released LC tasks in the HI mode, is set to zero (lines 14–16). Besides, there is a function (line 18) that checks whether the output of each task is ready. When ready, the task is removed from the core queue, and the generated dynamic slack is added to the slack array (lines 19–21). The learning process is conducted at the end of each hyper-period (lines 22–35). In this process, the number of dropped LC tasks in the HI mode and the accumulated dynamic slack is used to determine the state (line 23). As mentioned earlier, since the ϵ-greedy policy is used, if a random number is less than ϵ, a random action is selected (line 27, exploration phase of the learning process); otherwise, an action with the maximum value in the Q-table is chosen for that particular state (line 29, exploitation phase of learning process). Based on the chosen action, a new drop rate is determined for the task (line 31). Consequently, the reward function updates the Q-table (lines 32–33). Note that if an embedded system is kept running for a long time, although the system is in the exploitation phase over time, which uses the learned data, as can be illustrated from the policy, there is still a slight chance to learn if there is any pattern shift.

Algorithm 5.1 Proposed learning-based scheme at run-time

```
Input: Task Set, Core, Q-table
Output: QoS, Scheduled Tasks
procedure LEARNING-BASED QOS OPTIMIZATION ()
    SQ_t = 0;
    for t = 1 to Time do
        [Sys_MS, ReadyTaskQ] = TaskStatusCheck(Tasks)
        [Sch_tasks] = EDF-VD (ReadyTaskQ, Core)
        if Sys_MS == 1 then
            for each released LC Task_i do
                CountDrop_Task_i += 1;
                if mod(CountDrop_Task_i, δ_Task_i) == 0 then
                    NumDrop += 1;
                end if
            end for
        else
            for each Task_i do
                CountDrop_Task_i = 0;
            end for
        end if
        Flag_output=TaskOutputCheck(Tasks)
        if Flag_output == 1 then
            SlackQ= WCET_Task_i − Actualtime_Task_i
        end if
        if mod(t,HP)==0 then
            State= Deter-State (SlackQ,NumDrop)
            k= rand (1); //(0<k<1)
            //ϵ-Greedy Policy
            if k < ϵ then
                a_t = argrand (A_i)
            else
                a_t = argmax (s_t, A_i)
            end if
            Set the new task's drop-rate based on the action
            R = CompReward (U^MC(t)) // Eq. (5.4)
            Q(s_t, a_t) = Q(s_t, a_t) + α(R + γQ(s_{t+1}, a_{t+1}) − Q(s_t, a_t))//Eq. (3.21)
            SlackQ= 0; NumDrop = 0;
        end if
    end for
end procedure
```

5.3 Evaluation

The experiments are conducted on a Linux-based machine equipped with 1.4 GHz, a quad-core processor, and 16GB of memory. Since there are no real-life MC benchmarks to conduct the experiments, the major related studies have evaluated their proposed techniques by using synthetic task sets [2, 11–13]. However, in addition to synthetic task sets, we use various real-time tasks included in MiBench benchmark suite [14] for the evaluations. Besides, we analyze and compare the efficiency of proposed approaches against various methods [2, 12, 13, 15] in terms of schedulability, run-time QoS improvement, and available free slack at the end of a hyper-period. Our previous work (*FANTOM* [2]) uses the EDF-VD scheduling algorithm while dropping LC tasks in the HI mode according to the drop-rate parameter, without using the run-time adaptability. In [13](Liu+18), the QoS is improved by degrading the WCETs of LC tasks in the HI mode. In [15] (LRQ+14) and [12] (Hua+19), the run-time adaptability is employed by exploiting the accumulated dynamic slack. A dynamic reservation-based task scheduling algorithm has been presented in [15] (LRQ+14) to minimize the deadline miss rate by using the dynamic slack, which is generated by the early completion of HC tasks. In [12] (Hua+19), a FP-EDF is used, and the QoS is improved by precising the LC tasks' WCETs in the HI mode.

5.3.1 Evaluation with Real-Life Benchmarks

As mentioned, MiBench benchmark suite [14] has been used, which is dedicated to applications such as automotive, network, and telecommunications. More specifically, we consider the benchmarks of ‹edge›, ‹smooth›, ‹epic›, and ‹corner› as the LC tasks and ‹qsort›, ‹insertsort›, ‹matrixmult›, ‹dijkstra›, ‹bitcount›, and ‹FFT› as the HC tasks in our system evaluations. To achieve their execution times, these benchmarks have been executed on the ODROID XU4 hardware platform. We consider the utilization of the tasks in the [0.05,0.1] interval (to be able to execute more tasks in a core), and the period/deadline of the tasks is computed according to the utilization and WCET values [1, 2]. More detail on WCETs values has been reported in [16]. In addition, the values of the drop rate (δ) for the LC tasks are randomly generated between 1 and its maximum value (which has been set to four in our experiments), based on the uniform distribution. Table 5.2 represents the normalized Number of Deadline Misses (NDM), normalized to method of [13]), and the QoS of different methods. As shown in this table, at run-time, LIQUID has provided the maximum QoS and the minimum NDM compared to the other methods. Besides, since *FANTOM* [2] has also presented a design-time drop-aware approach to not drop LC tasks frequently when the system switches to the HI mode, the QoS has a high value (and low NDM value) compared to the other existing studies. Although the proposed techniques in [12, 15] are run-time approaches, which improve the QoS, they are not well designed to exploit better from run-time profit. However, since [15] uses the EDF-VD algorithm, which well utilizes the system's capacity compared to the other existing scheduling algorithms, e.g., FP

Table 5.2 NDM and QoS of different methods at run-time for a real task set

	LIQUID	*FANTOM*	[13]	[15]	[12]
NDM	0.009	0.010	1	0.341	0.999
QoS	99.67%	96.92%	64.33%	87.83	64.36%

algorithm, and also exploits the generated dynamic slack, it provides better results in terms of QoS and NDM in comparison with [12].

In addition, as mentioned in the previous chapter, we define a drop-rate value for each of the LC tasks in the HI mode. At run-time, these values are optimized based on the generated dynamic slacks. In our experiments, the drop-rate values for four LC tasks have been updated from $\{3, 3, 3, 3\}$ to $\{5, 6, 7, 7\}$, which has led to QoS improvement at run-time.

5.3.2 Evaluation with Synthetic Task Sets

Now we investigate the efficiency of LIQUID and SOLID under the presence of synthetic task sets. To generate synthetic task sets, analogous to [1, 13, 17], we consider dual-criticality task sets that are generated for various system utilization bounds ($U_{bound} = max(U_{LC}^{LO} + U_{HC}^{LO}, U_{HC}^{HI})$). We randomly add tasks to the task set to increase U_{bound}, until it reaches a given threshold in the [0.05,1] interval, with steps of 0.05. Besides, the periods of tasks are selected in the range of [100, 900] ms. For each utilization threshold, 50 task sets are generated. Since the mode switching probability determines how often the system switches to the HI mode, and, therefore, impacts the speed of learning in this mode, in the conducted experiments, we have considered different ratios of $\frac{WCET_{HC}^{LO}}{WCET_{HC}^{HI}} = Rat_{HC}$, from the [0.2,0.8] interval. The timing overhead of task execution interruption in the EDF task scheduling algorithm is in the order of μs [18], which has been considered as part of the tasks' WCETs in experiments. In addition, the actual execution time of a task follows the normal distribution of which the mean and standard deviation are $\frac{2*WCET_{HC}^{HI}}{3}$ and $\frac{WCET_{HC}^{HI}}{12}$ [19].

5.3.2.1 Effects of System Utilization

Considering the requirement of service maximization in MC systems, which is represented with drop-rate parameters for LC tasks, in this section, we first discuss the task schedulability under different utilization bounds in Fig. 5.3. Then, we evaluate the LC tasks' QoS at run-time under different methods by varying the system utilization bound. Figure 5.4 illustrates the run-time QoS improvement under different scheduling algorithms. In this experiment, the number of LC and HC tasks in each task set is almost the same ($Prob(HC) \simeq 0.5$). In addition, we do not change the run-time behavior of HC tasks while varying the utilization by considering an almost constant ratio of low-to-high WCET for the HC tasks ($Rat_{HC} \in [0.4, 0.6]$). In other words, since varying Rat_{HC} impacts objective values, we chose one subinterval and keep it constant, when the approach is analyzed by varying other parameters.

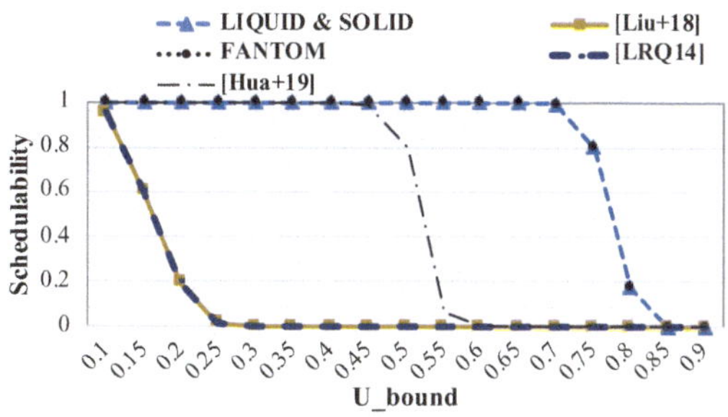

Fig. 5.3 Task schedulability of proposed method, FANTOM [2], [Liu+18] [13], [LRQ+14] [15], and [Hua+19] [12] by varying utilization bound

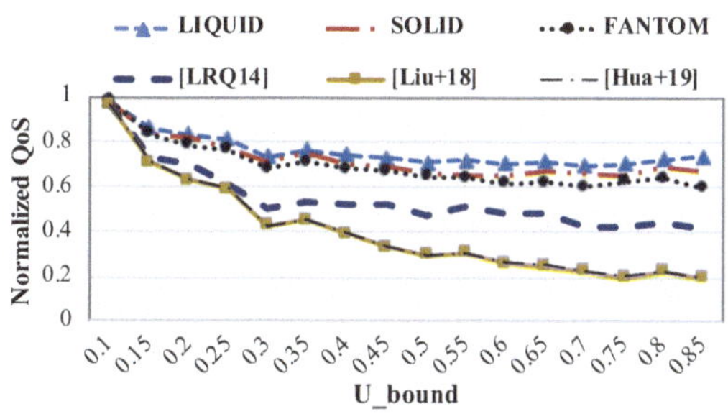

Fig. 5.4 Normalized QoS at run-time by varying utilization bound in proposed methods, FANTOM [2], [Liu+18] [13], [LRQ+14] [15], and [Hua+19] [12]

To evaluate the effects of varying utilization bound on the task schedulability at design-time, 1000 task sets are generated and evaluated. The task schedulability shows the ratio of task sets which are deemed as schedulable. Hence, a task set is schedulable if all HC and LC tasks can be executed correctly before their deadlines in the LO mode, and also all HC tasks and most LC tasks (based on their drop rate values) can be executed correctly before the deadlines in the HI mode. In fact, Eq. 5.2 and Eq. 5.3 must be satisfied in order to guarantee task schedulability. As shown in Fig. 5.3, since we have used the same design-time policy for the schedulability test as in [2], the results of these methods and our methods are the same. Besides, when the utilization is less than 0.7, LIQUID, SOLID, and FANTOM always schedule the tasks. This method could sometimes schedule the tasks as long as the utilization is smaller than 0.85. Moreover, LIQUID and SOLID provide a better task schedulability, compared to [12], due to using a different task scheduling approach than in [12], which is FP-EDF. Furthermore, this figure illustrates that the schedulability in methods of [13, 15] is worse than LIQUID. The main reason for this issue is that we prevent the frequent drop of LC tasks in the worst-case scenario (in the case of mode switches), while the other two methods may frequently drop LC tasks in the HI mode, and consequently, the task set would not be schedulable. Accordingly, the frequent LC tasks' dropping in other approaches may cause the system not to carry out its mission correctly [2].

Figure 5.4 shows the normalized QoS for different approaches. As shown, the amount of provided improvement is negligible in the low utilization bound. In addition, the QoS of the LC tasks is decreased by utilization increment in all methods. The reason is that utilization increment increases the number of both LC and HC tasks. Therefore, the system may switch to the HI mode more often. Although the LIQUID has more opportunity to learn due to more often mode switches and improve the learning process, all HC tasks' deadlines must be guaranteed in the HI mode, which causes more LC tasks to be dropped and, consequently, QoS reduction (QoS is the fraction of executed LC tasks before their deadlines to all LC tasks). However, due to the most favorable use of dynamic slacks at run-time, LIQUID has improved QoS better than the other methods. Besides, compared to the state of the art, *FANTOM* has provided more improvements in each utilization point due to maximizing the QoS at design-time. Besides, the QoS of the SOLID is better than *FANTON* due to using slack time at run-time to execute more LC tasks. However, in SOLID, since meeting the HC tasks' deadlines is guaranteed under any circumstances, and the dynamic slack is reclaimed carefully, it has less improvement than LIQUID. Since we evaluate the methods at run-time in terms of QoS, Li et al. [15] have exploited the dynamic slack to improve the QoS, while the other methods have almost the same behavior at run-time. As a result, the method of [15] has a better improvement compared to the results of [12, 13]. Note that the QoS is zero for $U_{bound} > 0.85$ in all methods due to the existing no schedulable task set under these methods when the utilization is more than 0.85. Since we guarantee that the LC tasks are not frequently dropped in the HI mode, more conditions must be checked to execute more tasks in the system, which leads the task sets to be unschedulable at high utilization.

5.3.2.2 Effects of HC Tasks' Run-time Behavior

Since we investigate the MC systems' run-time behavior, and the proposed method efficacy is influenced by how often the system switches to the HI modee, we vary the low WCET of HC tasks, determining the mode switching probability. In this regard, as part of our evaluations, we consider the low-to-high WCET ratio ($Rat_{HC} = \frac{WCET_i^{LO}}{WCET_i^{HI}}$) for the HC tasks in three different ranges of [0.2,0.4], [0.4,0.6], and [0.6,0.8]. Here, we assumed $U_{bound} = 0.75$, and the number of LC and HC tasks in each task set is almost the same ($Prob(HC) = 0.5$).

Figure 5.5 depicts the LC tasks' deadline miss rate for different approaches when varying low-to-high WCET ratio. The deadline miss rate is the ratio of the number of dropped LC tasks to the total number of tasks released in a time interval. Besides, the low-to-high WCET ratio increment means that the system switches less often to the HI mode due to having a high value of WCET for the HC tasks in the LO mode. The mode switching probability is decreased during run-time. As a result, it causes the system to be in LO mode most of the time, leading to fewer deadline misses. However, due to using the generated dynamic slack at run-time, the LIQUID reduces significantly the number of LC task drops compared to the other studies. Referring to the aforementioned reasons in the previous section, it has the same explanation for comparison with the results of the state of the art. Note that LIQUID may cause the HC tasks' deadlines to be missed. SOLID copes with this issue and executes the

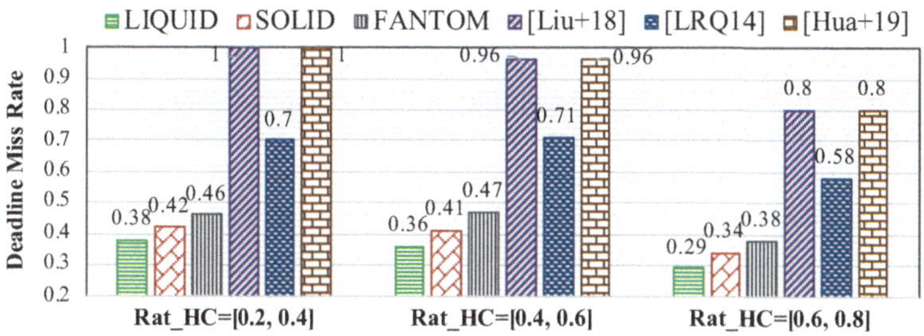

Fig. 5.5 Normalized deadline miss rate at run-time when varying low-to-high WCET ratio in proposed methods, FANTOM [2], [Liu+18] [13], [LRQ+14] [15], and [Hua+19] [12]

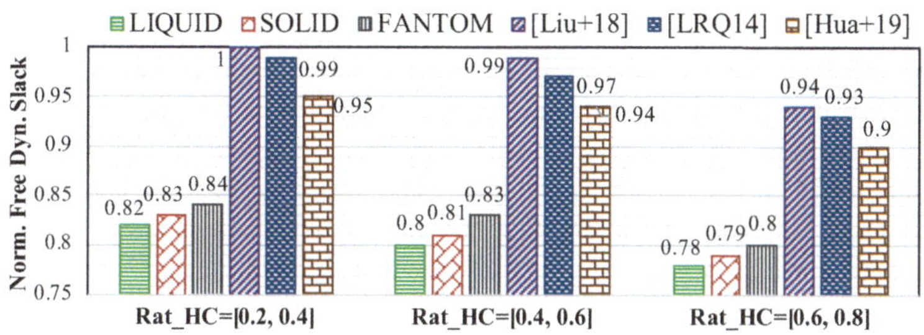

Fig. 5.6 Normalized unused free dynamic slack at run-time when varying low-to-high WCET ratio in proposed method, FANTOM [2], [Liu+18] [13], [LRQ+14] [15], and [Hua+19] [12]

LC tasks more, based on their newly learned drop-rate value if there is some slack generated by HC tasks' early finishes, to improve the QoS. In the end, the deadline miss rate is decreased under LIQUID by up to 47.87% and 32.45% on average, compared to the other works. In addition, SOLID could reduce the deadline miss rate by up to 43.47% (4.4% worse than LIQUID) and 28.33% (4.12% worse than LIQUID) on average.

Since we exploit the generated dynamic slack at run-time to prevent some LC tasks from dropping and consequently decrease the deadline miss rate, we investigate the amount of unused core utilization during a hyper-period. The amount of generated dynamic slack would be different in the HI mode while varying the low-to-high WCET ratio. As shown in Fig. 5.6, LIQUID can use more amount of dynamic slack, compared to the state of the art. Unlike *FANTOM*, which is a design-time approach, in LIQUID, we can use the dynamic slacks at run-time to improve the LC tasks' drop rates and then decrease the deadline miss rate. Furthermore, since the method of [13] is also a design-time approach, it could not use the free unused core utilization to improve the intended objective. The reason for having better core utilization usage in FANTOM, compared to [13], is the policy of executing more LC tasks in the HI mode (i.e., improving the QoS), and therefore, less unused core utilization is generated. Besides, [12] approach provides a better result in comparison with the technique of [15] due to its task scheduling policy, especially when the system switches to the HI mode. It seems that in [12], the system switches back sooner. Therefore, HC tasks are not executed up to their high WCET, which leads the system to spend more time in the idle mode during a hyper-period. Due to the possibility of HC tasks' deadline misses, SOLID handles it, and thus, less generated dynamic slack would be reclaimed. Based on our observations, exploiting the accumulated dynamic slack (generated at run-time) enables LIQUID and SOLID to reduce the free dynamic slack by 7.32% and 6.82%, on average, respectively, compared to the existing methods.

5.3.2.3 Impacts of Task Mixtures

We further evaluate the proposed approaches against the other methods under different HC task distribution variations. In this regard, Figs. 5.7 and 5.8 represent the deadline miss rate at run-time and the free dynamic slack in one hyper-period, respectively, when the HC tasks' utilization (i.e., more numbers of HC tasks, compared to LC tasks) to all of the generated tasks' utilization varies in three different ratio ranges of [0.2,0.4], [0.4,0.6], and [0.6,0.8]. Besides, in this part of our simulation, we assume $U_{bound} = 0.75$ and $Rat_{HC} = 0.75$.

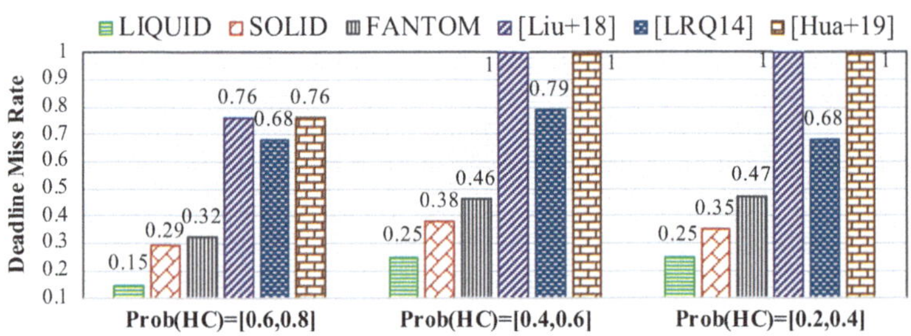

Fig. 5.7 Normalized deadline miss rate at run-time when varying task mixtures in proposed method, FANTOM [2], [Liu+18] [13], [LRQ+14] [15], and [Hua+19] [12]

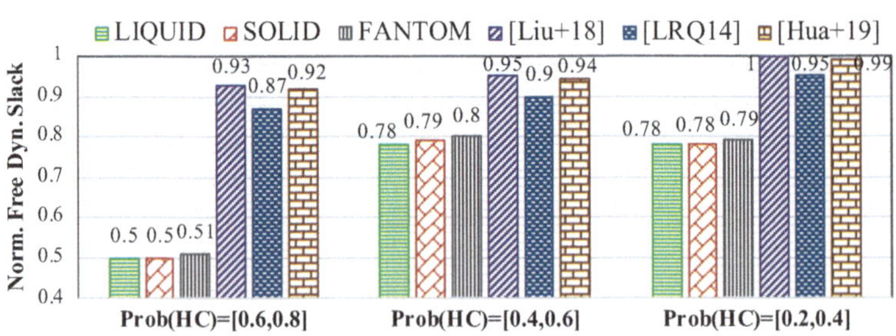

Fig. 5.8 Normalized unused free dynamic slack at run-time when varying task mixtures in proposed method, FANTOM [2], [Liu+18] [13], [LRQ+14] [15], and [Hua+19] [12]

According to Fig. 5.7, when more HC tasks are scheduled in the system, the mode switching probability is higher. It causes the system to drop LC tasks due to mode switching to execute all HC tasks correctly before their deadlines. It helps LIQUID to accelerate the learning process due to the more frequent mode switches and significantly improve the QoS by dropping fewer LC tasks in the future. In addition, since there would be fewer LC tasks in the system (by increasing the number of HC tasks in the system, while the system's utilization is constant), fewer LC tasks are dropped at run-time by the proposed schemes, which improves the QoS. Figure 5.8 shows that there is less free dynamic slack at the end of the hyper-period while increasing the $Prob(HC)$ range. In fact, since fewer LC tasks are scheduled in the system by increasing the $Prob(HC)$, the generated dynamic slack in the HI mode has been used for fewer LC tasks to improve their drop-rate value and, consequently, reduce the deadline miss rate. According to Fig. 5.7, LIQUID (SOLID) can decrease the deadline miss rate by up to 54.15% (44.88%) and 40.52% (30.39%) on average, compared to the state of the art. In addition, LIQUID (SOLID) can exploit the dynamic slack (generated at run-time) by 12.35% (11.95%) on average, in comparison with the existing methods.

5.3.2.4 Investigating the LC Tasks' Drop-Rate Parameter

Now, we evaluate the effects of varying the task mixtures on the increment of LC tasks' drop-rate parameters. This parameter is increased by exploiting the dynamic slack through a learning algorithm employed in the proposed scheme at run-time. Varying $Prob(HC)$ determines the percent of existing HC/LC tasks in a task set. Therefore, as mentioned in the previous section, when there are more HC tasks in the system, the QoS will have a higher value. It means that there is more increment in the drop-rate parameter value of LC tasks, which causes to drop fewer LC tasks and consequently decreases the deadline miss rate. This fact can be observed in Fig. 5.9, in which the average increment in drop-rate parameter for $Prob(HC) = [0.2, 0.4], [0.4, 0.6]$, and $[0.6, 0.8]$ are 47.05%, 66.32%, and 73.21%, respectively.

5.3.3 Investigating the Timing and Memory Overheads of ML Technique

Although the RL technique has been reported to be lightweight and highly suitable for the systems, compared to other types of learning techniques [20], the main issues are its convergence and timing overhead. Accordingly, similar to other studies [5], we have reduced the feasible actions to reduce the complexity and convergence issues. In the following, we investigate the timing and memory overheads of the employed learning technique.

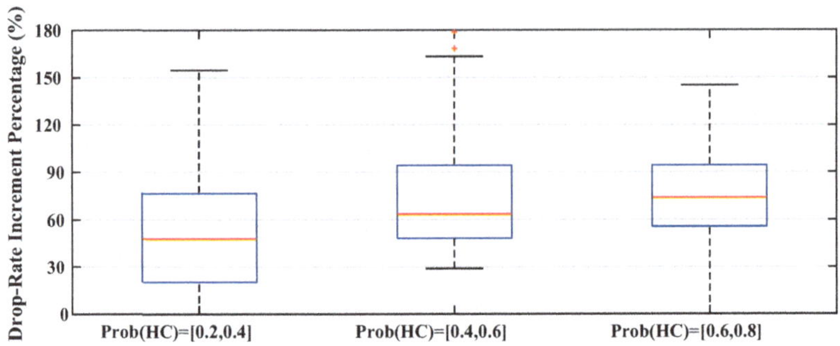

Fig. 5.9 Drop-rate increment percentage under the proposed approach by varying task mixtures

Consider a system with a single-core processor running an application with n tasks. To investigate the timing overhead of the learning process in each hyper-period, we analyze it on two systems, Intel® Core i7 processor and 2.5 GHz and ARM Cortex A-15 core and 2 GHz. From the complexity point of view, lines 22–35 of Algorithm 5.1 represent the learning process, in which for finding the maximum Q-value based on the obtained state (in a row of the table), a for-loop is used (line 29). Therefore, we can conclude that the complexity of the learning process depends on the number of actions in the Q-table ($O(A)$). According to our measurement at run-time, the maximum and average timing overhead in Intel core (ARM core) are 1.2 ms (4.1 ms) and 0.1 ms (0.53 ms), respectively. Since the maximum timing overheads are significant, to maintain the HC tasks' timeliness, we consider the learning process as a task with the WCET, equal to the maximum timing overhead and a period equal to hyper-period, while checking the task schedulability at design-time.

Furthermore, we need to clarify the amount of required memory space for storing the Q-table in terms of memory overhead. Accordingly, we store a two-dimensional array with $size(S)$ rows and $size(A)$ columns, in which the rows and columns show the states (S) and actions (A), respectively. Since the value of a table cell is in the range of $[-100, 100]$, it is required to consider at most 8 bits for storing each cell. As a result, we need $size(A) \times size(S) \times 8$ bits to store the Q-table. For an application with 30 tasks, the amount of required memory space for saving the Q-table with 100 states would be $30 \times 100 \times 8\text{bits} = 3\,\text{kB}$.

5.4 Conclusions

In this chapter, we proposed a novel approach, a learning-based drop-aware task scheduling mechanism, to reduce the deadline miss rate at run-time, with the aim of providing higher QoS. To achieve this goal, the dynamic slack is exploited at run-time, and since the system is unaware of the amount of generated dynamic slack in advance, the proposed scheme introduces an adaptive LC task dropping technique that uses an ML technique to exploit the slack and increase the survivability of LC tasks. Based on an extensive set of experiments, the proposed schemes can decrease the deadline miss rate by up to 51.78% and 31.32% on average and also exploit the accumulated dynamic slack generated at run-time by 9.84% more on average, compared to the current works. The proposed learning approach was analyzed regarding run-time timing overhead to ensure that there is no effect on missing the task deadlines. Although the timing overhead has been considered, it still has a significant value for embedded real-time systems, which is viewed as a limitation/drawback of the proposed scheme. Another limitation of the proposed method is the large exploration time of the learning process. Since the parameters would be updated through the learning process at the end of each hyper-period, the proposed method does not apply to applications with a large hyper-period. As an extension of this approach, if an MC system with more than two criticality levels is considered, there would be a challenge in how accumulated dynamic slack can be employed for improving the QoS of different criticality-level tasks, for example, for levels B, C, and D. Defining a threshold level of QoS for each criticality level might be fruitful in deciding how generated dynamic slack can be employed for each criticality level and improve their QoS.

Hitherto, we have proposed approaches to designing the MC systems with the aim of QoS improvement through application analysis. In these approaches, QoS has been improved through well-adjusted WCETs and task dropping analysis. Since the proposed approaches in Chaps. 3–5 can be applied to MC systems regardless of what hardware is used, i.e., single-/multi-core processors, we considered single-core processors as the hardware input. As mentioned in Chap. 1, in most embedded real-time applications, there are various tasks with different functionality, like controlling a device. Generally, these tasks have a common goal, like controlling autonomous driving, which may make them dependent. In other words, executing a task may depend on the complete execution of one/some of the other tasks. Therefore, since there are many tasks in these embedded real-time systems to be executed on a single platform, we can utilize the multi-core processors to execute the tasks in parallel to cope with the high-performance demands and improve the QoS. However, in MC hardware design, high-power consumption due to activating most of the cores may cause the systems to be more susceptible to failures and instability, which is not acceptable for MC systems, and it may cause catastrophic consequences. Therefore, the multi-core system's power and maximum temperature management must be considered while designing MC systems.

Although the proposed approaches in the last three chapters can be extended by considering the feature of multi-core platforms, the ability to execute tasks in parallel on multi-core platforms has not been considered in these approaches, which allows the QoS to be improved more efficiently. Besides, high-power consumption and temperature can be one of the crucial issues in these multi-core platforms that must be considered while designing such MC systems by the proposed approaches in order to guarantee the constraints like real-timeliness, safety, or management of the objectives like QoS.

To this end, in the following two chapters, we focus on MC hardware design and propose QoS-aware approaches to map and schedule the dependent MC tasks on multi-core platforms while reducing the power consumption and maximum temperature.

References

1. S. Baruah et al. "The preemptive uniprocessor scheduling of mixed-criticality implicit-deadline sporadic task systems". In: *Proc. of Euromicro Conference on Real-Time Systems (ECRTS)*. 2012, pp. 145–154.
2. B. Ranjbar et al. "FANTOM: Fault Tolerant Task-Drop Aware Scheduling for Mixed-Criticality Systems". In: *IEEE Access* 8 (2020), pp. 187232–187248. https://doi.org/10.1109/ACCESS.2020.3031039.
3. Zheng Li and Shuibing He. "Fixed-Priority Scheduling for Two-Phase Mixed- Criticality Systems". In: *ACM Transactions on Embedded Computing Systems (TECS)* 17.2 (2018), pp. 1–20.
4. B. Ranjbar et al. "Toward the Design of Fault-Tolerance- and Peak-Power-Aware Multi-Core Mixed-Criticality Systems". In: *arXiv preprint arXiv:2105-.07739* (2021).
5. Sai Manoj Pudukotai Dinakarrao et al. "Application and thermal-reliability-aware reinforcement Learning based multi-core power management". In: *ACM Journal on Emerging Technologies in Computing Systems (JETC)* 15.4 (2019), pp. 1–19.
6. H. Huang et al. "Autonomous Power Management With Double-Q Rein-forcement Learning Method". In: *IEEE Transactions on Industrial Informatics (TII)* 16.3 (2020), pp. 1938–1946.
7. D. Biswas et al. "Machine learning for run-time energy optimisation in many-core systems". In: *Proc. on Design, Automation & Test in Europe Conference Exhibition (DATE)*. 2017, pp. 1588–1592.
8. Hang Su, Dakai Zhu, and Daniel Mossé. "Scheduling algorithms for elastic mixed-criticality tasks in multicore systems". In: *Proc. on Embedded and Real-Time Computing Systems and Applications (RTCSA)*. 2013, pp. 352–357.
9. Dakai Zhu and Hakan Aydin. "Reliability-Aware Energy Management for Periodic Real-Time Tasks". In: *IEEE Transactions on Computers* 58.10 (2009), pp. 1382–1397. https://doi.org/10.1109/TC.2009.56.
10. Hang Su and Dakai Zhu. "An elastic mixed-criticality task model and its scheduling algorithm". In: *Proc. on Design, Automation & Test in Europe Conference & Exhibition (DATE)*. IEEE. 2013, pp. 147–152.
11. L. Sigrist et al. "Mixed-criticality runtime mechanisms and evaluation on multicores". In: *Proc. IEEE Real-Time and Embedded Technology and Applications Symposium (RTAS)*. 2015, pp. 194–206.
12. Lin Huang et al. "Improving QoS for global dual-criticality scheduling on multiprocessors". In: *Proc. of IEEE International Conference on Embedded and Real-Time Computing Systems and Applications (RTCSA)*. IEEE. 2019, pp. 1–11.
13. D. Liu et al. "Scheduling Analysis of Imprecise Mixed-Criticality Real-Time Tasks". In: *IEEE Transactions on Computers (TC)* 67.7 (2018), pp. 975–991.
14. M. R. Guthaus et al. "MiBench: A free, commercially representative embedded benchmark suite". In: *Proc. IEEE International Workshop on Work-load Characterization. WWC-4*. 2001, pp. 3–14. https://doi.org/10.1109/WWC.2001.990739.
15. Z. Li, S. Ren, and G. Quan. "Dynamic Reservation-Based Mixed-Criticality Task Set Scheduling". In: *Proc. of IEEE Intl. Conf. on High Performance Computing and Communications, IEEE Intl. Symp. on Cyberspace Safety and Security, IEEE Intl. Conf. on Embedded Software and Syst (HPCC,CSS, ICESS)*. 2014, pp. 603–610.

16. Behnaz Ranjbar et al. "BOT-MICS: Bounding Time Using Analytics in Mixed-Criticality Systems". In: *IEEE Transactions on Computer-Aided Design of Integrated Circuits and Systems (TCAD)* 41.10 (2022), pp. 3239–3251. https://doi.org/10.1109/TCAD.2021.3127867.
17. Z. Guo et al. "Uniprocessor Mixed-Criticality Scheduling with Graceful Degradation by Completion Rate". In: *Proc. on IEEE Real-Time Systems Symposium (RTSS)*. 2018, pp. 373–383.
18. Björn B. Brandenburg, John M. Calandrino, and James H. Anderson. "On the Scalability of Real-Time Scheduling Algorithms on Multicore Platforms: A Case Study". In: *2008 Real-Time Systems Symposium*. 2008, pp. 157–169. https://doi.org/10.1109/RTSS.2008.23.
19. Behnaz Ranjbar et al. "Power-Aware Runtime Scheduler for Mixed-Criticality Systems on Multicore Platform". In: *IEEE Transactions on Computer-Aided Design of Integrated Circuits and Systems (TCAD)* 40.10 (2021), pp. 2009–2023. https://doi.org/10.1109/TCAD.2020.3033374.
20. S. Pagani et al. "Machine Learning for Power, Energy, and Thermal Man- agement on Multicore Processors: A Survey". In: *IEEE Transactions on Computer-Aided Design of Integrated Circuits and Systems (TCAD)* 39.1 (2020), pp. 101–116.

Chapter 6
Fault-Tolerance- and Power-Aware Multi-core Mixed-Criticality System Design

In this chapter, we employ the hardware platform features in designing the MC systems in order to benefit from the parallel execution of tasks in multi-core platforms in improving the objectives, like QoS, while meeting the real-time constraints and safety requirements of tasks. To this end, we propose a method that exploits a tree of schedules for MC tasks (the tasks are dependent on this chapter). The proposed technique generates a tree of schedules offline (at design-time) considering all possibilities of fault-occurrence scenarios in different tasks (including both LC and HC tasks) and HC task overrun. At run-time, when an HC task overruns or a fault occurs in an LC or HC task, the scheduler chooses the proper schedule from the tree to tolerate the faults (by using the task re-execution technique) or manage the system mode switches with low overheads. The main goal of this proposed method is to improve the LC tasks' QoS in the HI mode while all HC tasks meet their deadlines. As a result, by generating the schedule tree and exploiting it at run-time, the LC tasks' QoS is maximized, while the occurrence of possible faults is tolerated. Since tasks can overrun and a fault can occur at any time but occasionally, using a single task's mapping and scheduling to guarantee the correct and on-time execution of all HC tasks without power constraint violation leads to inefficient utilization of resources. Thus, the proposed method manages peak power and temperature to prevent hotspots in homogeneous multi-core platforms.

To the best of our knowledge, this work is the first work to study the QoS-aware scheduling problem for fault-tolerant MC systems with peak power and thermal consideration. We summarize the main contributions of this work as follows:

- Proposing a tree generation approach for MC systems, based on all possibility of fault-occurrence scenarios and criticality mode changes
- Offline QoS-aware task mapping and scheduling to guarantee the correct execution of most LC tasks in the HI mode

B. Ranjbar et al., *Quality-of-Service Aware Design and Management of Embedded Mixed-Criticality Systems*, https://doi.org/10.1007/978-3-031-38960-3_6

- Peak power-aware task mapping and scheduling in multi-core MC systems for both LO mode and HI mode
- Reducing the run-time timing overheads by generating all schedules at design time and exploiting them at run-time

In this chapter, at first, a motivational example is presented in Sect. 6.1 for a better understanding of the problem and the proposed solution. Then, the design methodology and proposed method are explained in detail in Sects. 6.2 and 6.3, respectively. Finally, we analyze the experiments and conclude the chapter in Sects. 6.4 and 6.5, respectively.

6.1 Problem Objectives and Motivational Example

The MC system is responsible for running an application with a set of periodic dependent tasks on a multi-core chip. In addition, the system might face up to k transient faults in one application period. Prior to motivating by an example and explaining the details of our novel scheduling approach, we define the constraints, and the objective function, as follows:

Deadline Constraint Each HC task τ_i must finish its execution ($FTime_i$) correctly before its deadline (d_i) in both LO mode and HI mode. In addition, all LC tasks should finish their execution before their deadlines in the LO mode:

$$\begin{aligned} &\forall \tau_i,\ \zeta_i = HC : FTime_i \leq d_i \\ &\forall \tau_i,\ \zeta_i = LC \quad and \quad Cr_L = LO : FTime_i \leq d_i \end{aligned} \tag{6.1}$$

Task Dependability Constraint Due to the precedence correlations between tasks, the start time of task τ_i (st_i) must be greater than the finish time of all its predecessor tasks ($Pr(\tau_i)$):

$$\forall \tau_i, \forall j \in Pr(\tau_i) \implies st_i \geq FTime_j \tag{6.2}$$

Mapping Constraint A task (τ_i) can only be executed on a single core in each time slot. If X_{ij} denotes the mapping of task τ_i on core j, then:

$$\forall \tau_i,\ \sum_{j \in Cores} X_{ij} = 1 \tag{6.3}$$

Power Constraint The chip's overall power consumption must not violate the chip's TDP (TDP_{chip}) in any time slot:

$$\forall t \in timeslots : \sum_{j \in Cores} Pow_{jt} \leq TDP_{chip}, \tag{6.4}$$

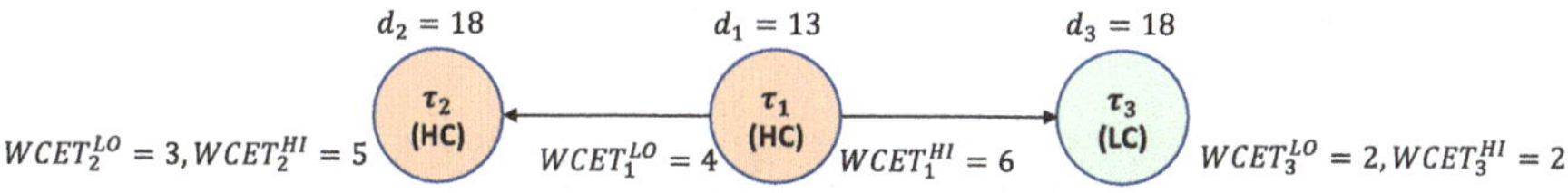

Fig. 6.1 Task graph of an application with three tasks used in the example

where Pow_{jt} represents the power consumption of core j in time slot t.

When the system switches to the HI mode, the system drops some LC tasks to meet the timing constraints, which degrades the QoS of the system:

$$Cr_L = HI : \quad QoS = n_L^{succ} / n_L \tag{6.5}$$

The problem is how to map and schedule dependent MC tasks of application A on the system's cores to satisfy the aforementioned constraints (timing and peak power) and QoS of the system. In this chapter, we propose a heuristic method to solve this NP-hard problem [1].

To better understand the problem, Fig. 6.1 shows an application with three dependent tasks, where tasks τ_1 and τ_2 have HC and task τ_3 has LC. Deadline, low WCET, and high WCET of each task are presented in the figure, and the system takes 1ms to discard the output of a faulty task ($\mu = 1$). Hence, the periods of all tasks are the same and equal to *18*ms. For the sake of simplicity, we considered that the application runs on a single-core processor, and up to one fault may occur during the execution of the application ($k = 1$). So, the system cannot execute multiple tasks simultaneously on different cores to violate TDP (we will discuss TDP challenge later). The scheduling algorithm in the LO mode executes tasks τ_1, τ_2, and τ_3, respectively. The schedulability test [2] shows that in the LO mode, all three tasks can be executed even in the case of fault occurrence. In other words, the total CPU utilization for application A is less than one ($U_A \leq 1$). If we consider the HI mode, if an LC task τ_3 is executed in addition to the two HC tasks in the case of fault occurrence, the system becomes overloaded (U_A^{HI}>1) and the three tasks cannot be scheduled. However, if we drop some LC tasks in the HI mode (τ_3 in this example) to guarantee the correct execution of HC tasks, then the computation demand requested by tasks is less than one and can be scheduled before their deadline. As a result, the utilization of this example for both LO mode and HI mode with the probability of one fault occurrence is computed as follows, in which just HC tasks are considered to be executed in the HI mode:

$$
\begin{aligned}
U_A &= MAX(U_A^{LO}, U_A^{HI}) \leq 1 \\
U_A^{LO} &= \left(\sum_{i \in \{1-3\}} \frac{WCET_i^{LO}}{T_A} \right) + \frac{k(\max_{i \in \{1-3\}}(WCET_i^{LO}) + \mu)}{T_A} \\
&= \frac{4}{18} + \frac{3}{18} + \frac{2}{18} + \left(\frac{4+1}{18} \right) = \frac{14}{18} < 1 \\
U_A^{HI} &= \left(\sum_{i \in \{1,2\}} \frac{WCET_i^{HI}}{T_A} \right) + \frac{k(\max_{i \in \{1,2\}}(WCET_i^{HI}) + \mu)}{T_A} \\
&= \frac{6}{18} + \frac{5}{18} + \left(\frac{6+1}{18} \right) = \frac{18}{18} \leq 1
\end{aligned} \tag{6.6}
$$

However, for this example, 14 different scenarios could happen during the execution of the application because the time of the fault and task overruns are unknown. Table 6.1 shows all these scenarios and the execution time of the system whether it drops LC task or not. In ten scenarios (S_5 to S_{14}), an HC task overruns, and it shows the system is in the HI mode. However, as shown in Table 6.1, only in two scenarios the system fails to execute all tasks (HC and LC tasks) before the deadline (S_7 and S_8). The reason is that the system has switched to the HI mode in these scenarios and also a fault has occurred. It causes the system to be overloaded and the computation demand for executing all tasks becomes more than one ($U_A > 1$). Therefore, the LC task τ_3 would be dropped. Although there are some scenarios such as S_{12} to S_{14}, that the system is in the HI mode, we schedule the LC task in this mode to improve the QoS. As shown in Table 6.1, the start time of τ_3 in S_{12} (S_{14}) is 16 (13), and since the WCET of the LC task is 2, then it can be executed before the application deadline ($d_A = 18$). This example clearly shows that all situations should be considered in an MC system. Therefore, the system is analyzed in detail at design-time, and then, the proper schedule is exploited in the online phase to minimize the drop ratio of LC tasks and enhance the QoS.

Figure 6.2 shows the tree for the application task graph presented in the motivational example, which is constructed in the offline phase of our proposed approach. At run-time, the system starts each period with S_1 (the scheduling in the root of the tree), which corresponds to the scenario where no fault occurs and no HC tasks overruns. If an HC task overruns (for instance, task τ_1), the system searches through the children of the current node (S_1), finds the appropriate task mapping and scheduling, and continues the execution based on the new schedule (S_6 in this case). After that, if the error detection unit detects a fault at the end of a task execution (for instance, task τ_2), the system searches through the children of the current node (S_6), finds the appropriate scenario, and continues the execution based on the new task mapping and scheduling (S_8 in this case).

Table 6.1 All possible scenarios of executing the task graph presented in Fig. 6.1 on a single-core processor

	S_1	S_2	S_3	S_4	S_5	S_6	S_7	S_8	S_9	S_{10}	S_{11}	S_{12}	S_{13}	S_{14}
Overrun	–	–	–	–	τ_1	τ_2	τ_1* ①	$\tau_1$①	$\tau_1$①	$\tau_2$①	$\tau_2$①	$\tau_1$②	$\tau_2$②	$\tau_2$②
Fault	–	τ_1	τ_2	τ_3	–	–	$\tau_1$②	$\tau_2$②	$\tau_3$②	$\tau_2$②	$\tau_3$②	$\tau_1$①	$\tau_1$①	$\tau_2$①
Finish time	9	14	13	12	13	11	**20*****	**19*****	16	17	12	18	16	15
Exe time (drop τ_3)	7	12	11	10**	11	9	18	17	14**	15	10**	16	14	13

* The circles shows the order of the occurrence of fault and task overrun

** In these cases, the system executes τ_3 and detects a fault occurs, but does not re-execute the task (drops the task)

*** In these cases, the system must drop task τ_3 to execute all HC tasks before their deadlines

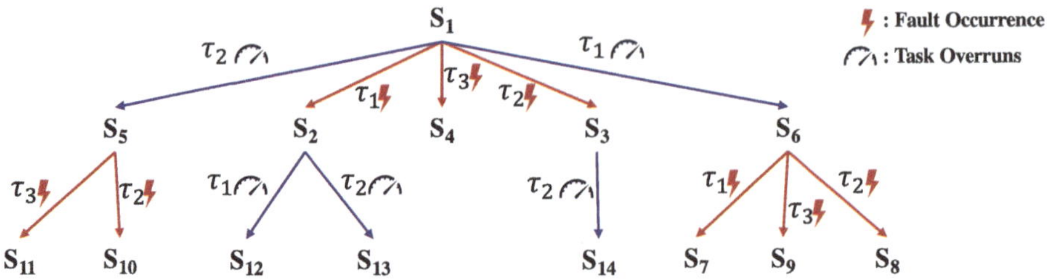

Fig. 6.2 The tree constructed by our method for the task graph of Fig. 6.1

It is important to mention that each schedule has a different start time, system mode, the expected number of faults, and task set. Furthermore, the scheduling of child nodes must be compatible with the scheduling of their parent, so the system can change the schedule of tasks without any conflicts. For instance, assume that in S_1, the task execution order is τ_1, τ_2, and τ_3. When the system employed S_4, it implies that τ_1 and τ_2 are completed successfully, the system is in the LO mode, and a fault is detected at the end of the first execution of τ_3. So, the S_4 should schedule tasks based on this information.

6.2 Design Methodology

The fault-tolerance and peak power-aware task mapping and scheduling method consists of two phases: design-time and run-time. In this chapter, we propose a design-time approach to be used at run-time for objective management. For the sake of completeness, we provide a brief overview of how to use the schedule tree in the run-time phase, which is generated at design-time. Figure 6.3 shows an overview of the design methodology. In the design-time phase, there are three functions that are used to generate the schedule tree, *MakeTreeRec*, *Schedule*, and *MapSch*. Section 6.3 provides details of generating the tree, these functions, and how we manage the peak power consumption. All scenarios are stored in memory to be used in the run-time phase. At run-time, an application follows the presented task mapping and scheduling at the root of the tree. In the case of fault occurrence or mode switching, the appropriate task mapping and scheduling for the remaining un-executed tasks is fetched from memory. After fetching, mapped tasks based on the previous scenario are remapped based on the new scenario, and the system continues its operation. In the following subsection, we explain the design-time phase of our proposed method.

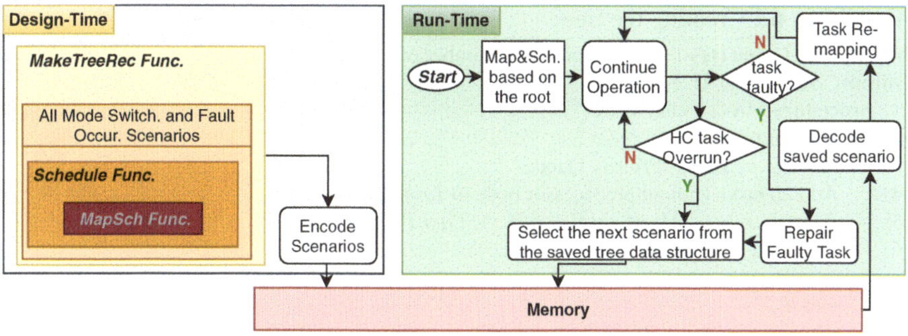

Fig. 6.3 Design methodology

6.3 Tree Generation and Fault-Tolerant Scheduling and Mapping

As we discussed, multiple scenarios might happen during the execution of an instance of the application, where, in most of these scenarios, the system can execute all or most of LC tasks without violating HC tasks' deadline. To this end, the proposed approach of this work considers a different mapping and scheduling for each scenario to handle HC tasks' deadlines, faults, and peak power violations while minimizing the number of dropped LC tasks in the HI mode. In the run-time phase, in general, the system is unaware of tasks that might overrun or a fault that occurs, so the system cannot select the proper schedule in advance. Therefore, this work employs a tree data structure in the offline phase to organize the mapping and scheduling of tasks for all scenarios, corresponding to each HC task overrunning, and/or up to k fault occurrence during each period. Now, we explain how the scheduling tree is generated.

6.3.1 *Making Scheduling Tree*

The main function of creating the tree (Φ) is outlined in Algorithm 6.1. At first, we define a priority queue called *TaskPQ* (line 3), which considers tasks' release time as the priority. The release time of a task is the time when all its predecessor tasks have finished their execution (presented in Sect. 2.1.1.1). Then, the algorithm enqueues all tasks without any predecessor to the *TaskPQ* with a key equal to 0, because they are released at the beginning of the period (line 4). Each node of the tree represents a particular scenario, and it has two attributes called *sch* and *childs*. For instance, in the root scenario, the system is in the LO mode, and no fault occurs in the entire period. *sch* is the proper mapping and scheduling of tasks for that scenario, and *childs* is a list of children nodes of the current node.

Algorithm 6.1 Creating the tree

```
Input: Task Graph (G_T), List of Cores (c), Number of Faults (k).
Output: Scheduling Tree (Φ).
procedure MAKETREEMAIN ()
    Φ ← Empty Tree
    TaskPQ ← Empty Priority Queue
    Add all tasks without predecessor node to TaskPQ with key 0
    Φ[root].sch ← MapSch(G_T, c, 0, ∅, TaskPQ)
    if Φ[root].sch=un-scheduled then
        return ∅
    end if
    Φ[root].childs ← MakeTreeRec(G_T, c, Φ[root].sch, 0, k, LO)
    if Φ[root].childs=un-scheduled then
        return un-scheduled
    end if
    return Φ
end procedure
function MAKETREEREC (Task Graph (G_T), List of Cores (c), Parent Schedule(Sch), Time (T), Number of Faults (k), Mode of the System (Mode))
    SchList ← Empty list of nodes
    if Mode = LO then
        //HChild nodes
        Tasks ← List of unfinished HC tasks in time T.
        G_H ← G_T With High WCET.
        for each τ in Tasks do
            T_tmp ← Finish time of τ in sch + T_sw.
            S ← New node
            S.sch = Schedules (G_H, c, T_tmp, sch, TaskPQ)
            if S.sch = un-scheduled then
                return ∅.
            end if
            S.childs ← MKTreeRec(G_H, c, S, T_tmp, k, HI)
            if S.childs = ∅ then
                return ∅.
            end if
            Add S to SchList
        end for
    end if
    if k > 0 then
        Tasks ← List of Unfinished Tasks in T.
        for each τ in Tasks do
            T_tmp ← Finish Time of τ_i in sch.
            S.sch = Schedules(G_T, sch, Time)
            if S.sch = un-scheduled then
                return un-scheduled.
            end if
            S.childs ← MKTreeRec(G_T, c, S, T_tmp, k − 1, Mode)
            if S.childs = ∅ then
                return ∅.
            end if
            Add S to SchList
        end for
    end if
    return SchList
end function
```

Algorithm 6.2 Schedule procedure

Input: Task Graph (G_T), List of Cores ($Cores$), Time (T), Parent Schedule (Sch_{par}), Ready Task Priority Queue ($TaskPQ$).
Output: Schedule (Sch)
1: **procedure** SCHEDULES ()
2: $T_r \leftarrow$ List of Tasks from G that all predecessors has started executing before time T.
3: Add T_r Tasks to a Priority Queue (*TaskPQ*).
4: $Sch \leftarrow Sch_{par}[0 - T]$
5: $Sch \leftarrow MapSch(G_1, Cores, T, Sch, TaskPQ)$
6: **if** Sch = un-scheduled **then**
7: Find a LC Task with Largest execution time which has not started in T and remove it from G then **goto** line 2.
8: **if** Cannot find any LC task **then**
9: **return** un-scheduled.
10: **end if**
11: **end if**
12: **return** Sch
13: **end procedure**

The algorithm calls *MapSch* function (Algorithm 6.3, which is discussed later in this section) to schedule task for the root node (line 5). The algorithm returns *un-scheduled* if *MapSch* function cannot find any feasible schedule with no task dropping and violating the TDP constraint (lines 6–8). Otherwise, the algorithm in line 9 continues to create the rest of the tree recursively by calling *MakeTreeRec* function (which is presented in lines 15–50 of this algorithm). If task scheduling is feasible in all possible scenarios, *MakeTreeRec* function returns a list of child nodes, and the algorithm returns the tree (Φ); otherwise, the algorithm returns *un-scheduled*, which means it could not find a feasible solution (lines 10–13).

The *MakeTreeRec* function in Algorithm 6.1 recursively creates the tree. Each node in the tree might have two types of child nodes. The first type of child node (*HChild*) has a scenario similar to their parents, except that one of the HC tasks overruns. Therefore, if the system is in the LO mode, any unfinished HC task might overrun and change the system's mode. To this end, first, *MakeTreeRec* function collects all unfinished HC tasks and creates an HC task graph (G_H) by changing the WCET of all tasks to high WCET (lines 18–19). Then, for each unfinished HC task, the function considers the scenario that the task overruns and schedules the task by calling *Schedules* function, presented in Algorithm 6.2. If the *Schedules* function finds feasible scheduling, the algorithm recursively creates a tree for this node, where the system is in the HI mode, and up k faults may occur on the remaining tasks by calling *MakeTreeRec* function (lines 20–32). It is important to mention that switching to the HI mode has nonzero timing overhead (T_{sw}) in realistic systems [3], but it is insignificant in comparison with tasks' WCETs (line 21). The second type of child node (*FChild*) has a scenario similar to their parents, except for one fault that occurs during the execution of one of the remaining tasks. The system can tolerate up to k faults in a period. If less than k faults occur in a node scenario, a faulty execution of all remaining tasks needs to be considered. Therefore, a child node is generated for the faulty execution of each remaining task, and also for each child node, the algorithm recursively constructs a tree by calling *MKTreeRec* with *k-1* faults (line 34–48). Finally, if the algorithm finds a feasible solution for all scenarios, it returns the list of child nodes ($SchList$).

The *Schedule* function schedules tasks for each situation by calling *MapSch* function (Algorithm 6.3). If *MapSch* function fails to find a feasible solution to meet the deadlines of all tasks with respect to TDP constraint, *Schedules* function drops the largest LC task (in terms of WCET) and calls *MapSch* function again. The *Schedules* function repeats this procedure to find a feasible schedule. If *MapSch* function fails to find a feasible solution, and there is no more LC task to drop, *Schedules* function returns *un-scheduled* (lines 2–12 Algorithm 6.2). We will discuss the *MapSch* function in the next subsection. As we mentioned in this section, occurring faults and a criticality mode change generate different scheduling scenarios that correspond to a set of alternative schedules. These scenarios are stored in the memory of the system as a tree in the offline phase. At run-time, the system starts with the scheduling in the root node, which is for the scenario where no fault or overrun happens. After that, if a fault occurs or an HC task overruns, the system finds the appropriate scenario in the child nodes of the current node and changes the scheduling of the system to improve the number of executed LC tasks.

6.3.2 Mapping and Scheduling

In this section, we explain the proposed mapping and scheduling algorithm, which manages the peak power and hotspot distribution. It should be mentioned that low-power techniques, e.g., DVFS, cannot be easily used in the HI mode, especially when the system is in the overload situation due to the timing overhead. Therefore, we manage the peak power by finding the proper mapping and scheduling of tasks on free time slots of cores. This task mapping and scheduling is feasible if the system's power consumption never exceeds the TDP constraint, and all tasks finish their execution (even in the worst case) before their deadline. So, *MapSch* algorithm decides the time and core where each task should be executed.

Algorithm 6.3 outlines the pseudo-code of the *MapSch* algorithm. Tasks are mapped and scheduled up to time *T* based on the current node schedule. In the case of fault occurrence or mode switches at time *T*, this algorithm maps and schedules the rest of the tasks based on the new node schedule from time *T* to the end of the application period (*PERIOD*). The time is divided into a set of equal time slots (*TS*), and the scheduler will put tasks into cores only at the beginning of each time slot. In each time slot, at first, the algorithm sets an empty array for ready tasks, and then it extracts all elements of *TaskPQ*, where their key is equal to the current time slot. This means that all predecessor tasks of these ready tasks have finished their execution. If *TaskPQ* and *ReadyTasks* array are both empty, the algorithm returns the final scheduling (*Sch*) because it successfully schedules all tasks. If there is no

Algorithm 6.3 Mapping and scheduling pseudo-code

Input: Task Graph (G_T), List of Cores ($Cores$), Time (T), Scedule up to the Time (Sch), Ready Task Priority Queue ($TaskPQ$).

Output: Complete Schedule (Sch)

```
1: procedure MAPSCH ()
2:     for TS = T to PERIOD do
3:         ReadyTasks ← ∅
4:         Extract minimum element from TaskPQ and add it to ReadyTasks while key
            of each element is equal to TS and TaskPQ is not empty;
5:         if TaskPQ.empty() = true and ReadyTasks = ∅ then
6:             return Sch // Scheduling is done;
7:         end if
8:         if ReadyTasks = ∅ then
9:             continue // No new task is ready in this TimeSlot
10:        end if
11:        SortedTasks ← Sort (ReadyTasks, Desc);
12:        SortedCores ← Sort (Cores, Asc);
13:        for task in SortedTasks do
14:            for core in SortedCores do
15:                Time_tmp ← task_wcet
16:                count ← 0
17:                Sch_tmp ← Sch
18:                SysPow_tmp ← SysPow
19:                while Time_tmp > 0 do
20:                    if Sch_tmp(TS+count, core) is empty & SysPow_tmp(TS+count) +
                        task_pow <= TDP then
21:                        Sch_tmp(TS + count, core) = task
22:                        SysPow_tmp(TS + count) += task_pow
23:                        Time_tmp -= 1
24:                    end if
25:                    count += 1;
26:                end while
27:                if TS+count ≤ task_deadline then
28:                    Sch ← Sch_tmp
29:                    SysPow ← SysPow_tmp
30:                    SortedCores ← Sort (Cores, Asc);
31:                    task_sch ← true
32:                    break
33:                end if
34:            end for
35:            if task_sch == false then
36:                return un-scheduled
37:            end if
38:        end for
39:    end for
40: end procedure
```

ready task to be scheduled in the current time slot (the *ReadyTask* array is empty, but the *TaskPQ* is not empty), the algorithm moves to the next time slot (lines 2–10).

The algorithm sorts the ready tasks in descending order of their energy consumption (line 11). The energy consumption of each task τ_i (Eng_i) is calculated as follows:

$$Eng_i = Pow_i \times WCET_i \tag{6.7}$$

where Pow_i and $WCET_i$ are the maximum power consumption and the worst-case execution time of task τ_i. The maximum power of each task can be obtained by running benchmarks on a real platform. As mentioned in Sect. 2.1.4, the processor power consists of three components; when a task is run on a processor, the dynamic power is increased significantly compared to static and independent powers. Hence, in this work, we do not model the power; we measure the processor power when tasks are run on the real platform. More information about computing these values is given in Sect. 6.4.1. The system's power consumption must never exceed the TDP constraint to overcome the overheating problem [1]. To this end, we consider a constant power consumption for each task at design-time, which is equal to its maximum power consumption, to guarantee the meeting of TDP constraint in the worst-case scenario. In addition, energy increment leads to an increase in chip temperature [4, 5]. Thus, we map a task with more energy consumption to a core with less temperature. Then, the algorithm sorts the cores in the ascending order of their accumulated energy (line 12). A core has a higher priority for task assignment if it has less accumulated energy (i.e., tends to have a less temperature degree).

After sorting tasks and cores, the algorithm assigns tasks to the cores one by one (lines 13–38). So, for each task, the algorithm selects a core from the sorted list and schedules the task on the core's free slots (lines 19–26). The system's instantaneous power consumption must be less than the TDP constraint, so we consider an array called $SysPow$, which holds the maximum power consumption of the system in each time slot. The algorithm checks the $SysPow$ and TDP constraint before scheduling a task on a core (line 20). If the task is completed before its deadline, the algorithm updates the schedule (Sch), power array ($SysPow$), and scheduling status of the task ($task_{sch}$). It also sorts the cores again since the energy of one core has changed and starts to schedule the next task (lines 27–33). If the task does not meet its deadline on the core, the algorithm picks the next core and schedules it on that core. However, if the deadline of one task is violated in all cores, the algorithm fails to schedule tasks of the application in this scenario and returns "un-schedulable" (lines 36–37).

The example in Fig. 6.4 shows an MC application and how our method maps and schedules the tasks on three cores. Assume the TDP constraint is 1.6 watts. Figure 6.4b shows that the scheduling without our policy violates the TDP constraint, while the maximum power consumption of the task scheduling by considering our policy is below the TDP constraint (Fig. 6.4c). When the system completes task τ_1, three tasks (τ_2, τ_3, and τ_4) become ready to be executed. So, $Eng_{\tau_2} > Eng_{\tau_3} = Eng_{\tau_4}$ that shows τ_2 should be mapped to the core with less accumulated energy (E_c). In addition, as can be computed, $E_{C3} = E_{C2} > E_{C1}$. Therefore, according to the criticality level, we first map τ_2 and τ_3 on C_3 and C_2, respectively, and schedule them, and thereafter, τ_4 is mapped on C_1. This procedure will be the same for mapping and scheduling τ_5 and τ_6.

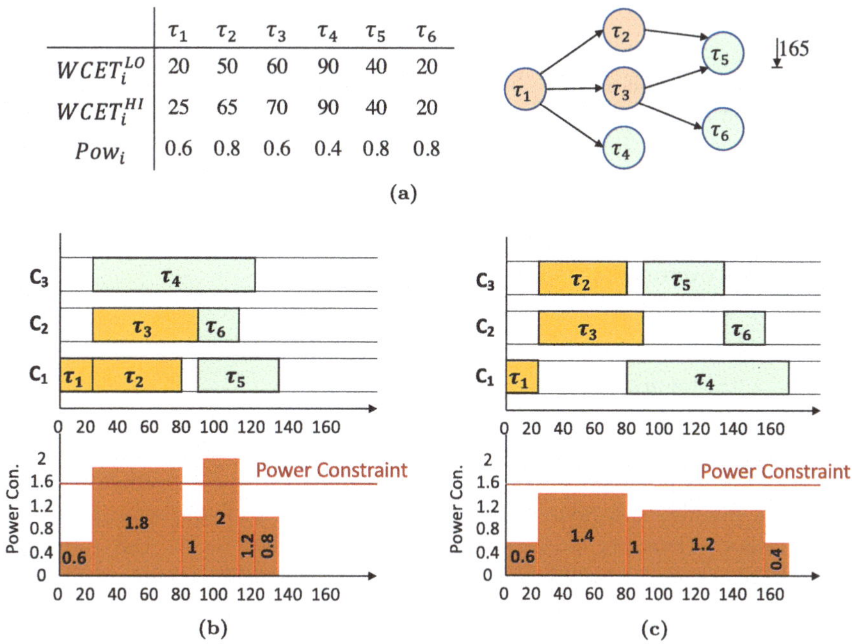

Fig. 6.4 Two different task scheduling scenarios. (**a**) An example of MC application. (**b**) Task scheduling without our policy. (**c**) Task scheduling with our policy

6.3.3 Time Complexity Analysis

In this section, we consider an m-core processor running an application with n tasks and k possible fault occurrences during the application's execution. At first, we describe the complexity of generating a tree as the main computation part of the proposed method.

When an HC task overruns, the system switches to the HI mode, and the scheduler considers the high WCET of remaining tasks until the complete execution of the application. So, in each execution, only one task may overrun, and the possible scenarios for overrun situations are equal to the number of HC tasks (n_H). Furthermore, we assume that up to k faults may occur during the execution of the application. For clarity, we first compute the number of different fault-occurrence scenarios for $k = 0$, 1, and 2. Then, we present a general formula to obtain the maximum number of possible scenarios (nodes of the tree).

- $k = 0$: In this case, there is one scenario for a situation where none of the tasks overruns (the root of the tree) and n_H (number of HC tasks in the graph) scenarios for situations where one of the HC tasks overruns (leaf nodes of the tree). Thus, the number of all schedules is:

$$T(k=0) = 1 + n_H \tag{6.8}$$

- $k \leq 1$: In this case, the tree has $T(k = 0)$ nodes to handle $k = 0$ scenarios, in addition to the nodes which contain scheduling for $k = 1$ scenarios. There are three possible situations for the scenarios where one fault occurs to a task (HC or LC task). There are n scenarios for situations when no HC task overruns, $n \times n_H$ scenarios for situations where the faulty task executes before an HC task overruns, and $n \times n_H$ scenarios for situations where the faulty task executes after an HC task overruns. Therefore:

$$\begin{aligned} T(k \leq 1) &= T(k=0) + n + 2 \times n \times n_H = 1 + n_H + n_H \times n + n \times (1 + n_H) \\ &= 1 + n_H + n_H \times n + n \times T(k=0) \end{aligned} \tag{6.9}$$

- $k \leq 2$: In this case, the tree has $T(k \leq 1)$ nodes to handle $k \leq 1$ scenarios, in addition to the nodes which contain scheduling for $k = 2$ scenarios. There are four possible situations for the scenarios where two faults occur to one or two task(s). There are n^2 scenarios for situations when no HC task overruns, $n \times n \times n_H$ scenarios for situations where the two faulty tasks are executed before an HC task overruns, $n_H \times n \times n$ scenarios for situations where the two faulty tasks are executed after an HC task overruns, And $n \times n_H \times n$ scenarios for situations where an HC task overruns in the middle of two faulty tasks. It is noteworthy to mention that there are scenarios in which both faults and overrun happen on a one HC task (similar to S7, S10, S12, and S14 in Table 6.1). Therefore:

$$\begin{aligned} T(k \leq 2) &= T(k \leq 1) + n^2 + 3 \times n_H \times n^2 \\ &= 1 + n_H + n_H \times n + n_H \times n^2 + n \times T(k \leq 1) \end{aligned} \tag{6.10}$$

Therefore, we can conclude that the maximum number of possible schedules by considering maximum k fault occurrence is:

$$T(k) = 1 + n_H \left(\sum_{i=0}^{k} (n^i) \right) + nT(k-1), T(0) = 1 + n_H \tag{6.11}$$

By solving Eq. (6.11), we can conclude that generating the tree is in order of $O(n^{k+2})$:

$$T(k) = n_H \times \left(\sum_{i=0}^{k} ((i+1) \times n^i) \right) + \frac{1 - n^{k+1}}{1 - n} \tag{6.12}$$

Please note that this value is a general upper bound for the generating tree algorithm, and for an actual task graph, the total number of scenarios is less than Eq. (6.11). The reason is that the total number of scenarios is presented with no awareness of the exact dependency between tasks to count the precise number of scenarios when a fault occurs and then a task overruns, or vice versa. For example, there are 14 different scenarios for two HC tasks, one LC task, and $k = 1$ for the task graph presented in Fig. 6.1, while $T(k)$ is equal to 18 in Eq. (6.11).

6.3.4 *Memory Space Analysis*

In this section, we discuss the memory space needed for storing the scheduling tree. For each scenario, we store two arrays with the size of the number of tasks. The first array determines the core assigned to each task, and the second array determines the start time of the tasks. In the first array, we denote that each task is mapped to which of c cores. So, each task requires $\log_2^c$ bits. Since we have n tasks in the application, the total memory space required for each scenario is $n \times \log_2^c$ bits. Considering the period of the application and the size of each time slot, the second array size is equal to $n \times \log_2^{period/timeslot}$. Therefore, the total amount of needed memory (bits) is:

$$Mem(n, c, k) = T(k) \times \left(n \times \left(\log_2^c + \log_2^{\frac{period}{timeslot}}\right)\right) \tag{6.13}$$

Assuming c and *period/timeslot* values are less than 2^{32}, the memory space needed for saving the scheduling tree for an application with 32 tasks and up to two possible fault occurrences in the worst-case scenario is less than 13 MB. It is noteworthy to mention that the scheduling tree can be stored in the FLASH or read-only memory of the system, and there is no need to load the whole tree to the RAM at run-time. In the case of fault occurrence or mode switching at run-time, the system discards the current schedule and loads the proper child node's schedule into the RAM. In this example, our approach occupies less than 2 KB of the RAM.

6.4 Evaluation

6.4.1 *Experimental Setup*

6.4.1.1 Application

For the experiments, we used both real-life and random applications to show our proposed approach's efficacy. To generate random task graphs, we used the tool presented by Medina et al. [6]. We generated applications with 30, 40, 50, and 100 tasks (n), where 20–50% of them are LC tasks. Another important parameter in a task graph is edge percentage (d) which shows the probability of having edges from one task to another task. We considered 1–20 % edge percentage in the experiments. Another important parameter that will be discussed in the experiments is the normalized system utilization U/c, where U is the utilization of the system considering the high WCET of each task and c is the number of cores. We considered different values of normalized system utilization in the range of (0,1] with the steps of 0.05. We also evaluate the proposed approach and other approaches for comparison, with a real-life application task graph, vehicle Cruise Controller (CC) [7], composed of 32 tasks, where 34% of them are LC tasks. In addition, the value of edge percentage for this CC application is 7%.

6.4.1.2 Hardware Platform

To evaluate our approach, we conduct the experiments and run the applications on a platform with 2, 4, 8, and 16 cores, which models ARM Cortex-A7 cores (c). The maximum number of transient faults that may occur during each application period (k) and the recovery overhead μ are considered 3 and 15 ms, respectively [8]. It is important to know that if λ and $time$ is the fault rate and application execution time, respectively, the minimum number of fault occurrence would be $\lambda \times time$. Therefore, k would not be much smaller or larger than $\lambda \times time$ [9]. If $\lambda = 10^{-6} fault/\mu s$, and $time = 10^3$ ms, then $\lambda \times time = 1$. As a result, since this fault rate is much higher than real fault rates, mentioned $10^{-12} fault/\mu s$ in [10], considering $k \leq 3$ is a reasonable fault-occurrence number during each application period. We use the HOTSPOT tool [11] to obtain the cores' temperature trace by exploiting the specific floorplan according to a real platform, ODROID XU3 board, which has four ARM Cortex-A7 cores, and the parameters used in [12]. In addition, we use the reported value in [3], to consider the timing overhead of mode switching for ARM Cortex processors. We considered the maximum reported overhead, which is $T_{sw} = 254\ \mu s$, in our experiments.

6.4.1.3 Peak Power Consumption

To determine a realistic power consumption for tasks, we ran several embedded benchmarks from the MiBench suite, such as automotive, network, and Telecomm., on the ARM Cortex-A7 core of the ODROID XU3 platform with maximum frequency at design-time. We monitor the power sensors continuously, and we set the worst measured power as the power consumption of tasks. In addition, we examined different scenarios of activating one core to all cores by running different benchmarks. Each benchmark is run 1000 times on a core, and we considered each task's power consumption between the minimum and maximum power values obtained from the platform. The measurement reports show that the power consumption of tasks is between 483 mW and 939 mW. In these experiments, the TDP value has been considered 85% of the maximum power that a chip can consume, which is used conventionally in embedded processors [13].

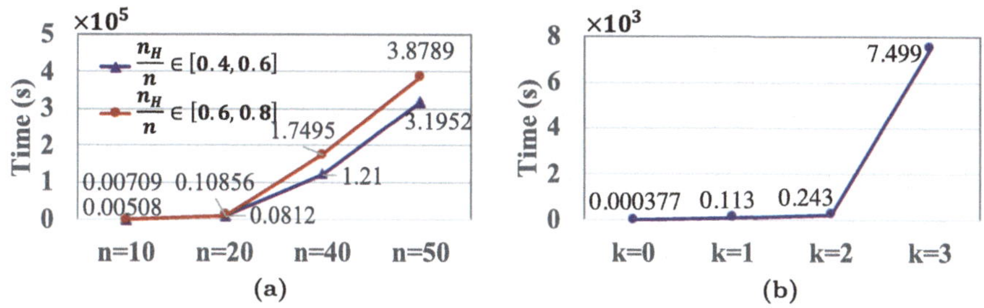

Fig. 6.5 Tree construction time for different application sizes and number of fault occurrence. (**a**) Number of tasks. (**b**) Number of fault occurrence

6.4.1.4 Comparison

We analyzed our proposed method and compared our experimental results to the results obtained by recent works that use the task graph model [6, 14, 15]. Socci et al. [14] ([Soc+15]) have proposed an online scheduling algorithm for an MC system where only HC tasks are executed in the HI mode. Medina et al. [6] ([MBP18a]) have considered a fault-tolerant MC system that generates two tables at design-time and uses them at run-time. Based on this work, we in [15] ([Ran+19]) have presented an online approach (will be presented in detail in the next chapter) to reduce the peak power and temperature by using the DVFS technique. Hence, due to the timing overhead of DVFS and increasing fault rate by changing the *V-f* levels [16], we cannot easily use this technique, especially in the HI mode.

6.4.2 *Tree Construction Time*

At first, we evaluate offline tree construction time by varying the parameters n and k, in Fig. 6.5. The tree's construction time is computed on a system with an Intel Core-i5 processor with 1.3 GHz clock frequency. Construction time depends on the number of faults and tasks. Figure 6.5a shows the effects of the number of tasks and the portion of HC tasks in each task set with $k = 3$. Besides, Fig. 6.5b shows the effects of the number of fault occurrences with $n = 20$. These figures depict that by increasing the number of faults or number of tasks, the tree generation's time is increased exponentially. Also, task sets with higher HC tasks have higher tree construction time. Although the offline tree construction time is relatively high for large applications, the online overhead is small and constant for all applications. It is noticeable that our method can generate each node of a tree in parallel to reduce the construction time. For example, if we have a system with four cores, the construction time is about four times faster than in a single-core system.

6.4.3 Run-Time Timing Overheads

In case of fault occurrence or mode switching, the system finds the proper schedule by moving to the child of the current node, which is responsible for the upcoming scenario. Each node of the tree stores two arrays with the size of $n \times (log_2^c + log_2^{\frac{period}{timeslot}})$ bits, where c and n are the number of cores and tasks, respectively. Thus, the switching time between the schedules consists of moving one level in the tree and retrieving the correct scheduling from memory, which is constant and negligible. We measured the schedule changing time at run-time on the ODROID XU3 platform; considering $c = 8$, and $n = 50$, it is almost 0.47 μs.

6.4.4 Peak Power Management and Thermal Distribution for Real-Life and Synthetic Applications

In this subsection, we analyze the approaches in terms of peak power and maximum temperature, by running two real-life applications: vehicle CC [7] and object detection function using LIDAR sensor in Autoware application [17].

Figure 6.6 shows the system's power traces of vehicle CC application [7] by our proposed approaches, the approaches proposed by Socci et al. [14], Medina et al. [6], and Ranjbar et al. [15]. Since [15] is the online approach to minimize the peak power while exploiting the same task mapping and scheduling of [6] at design-time, their power traces and thermal distributions are the same in the worst-case scenario of tasks' execution time and power consumption. In this part, to focus on the behavior of systems in the HI mode, we assumed no fault occurred during the application's execution. Socci's approach does not violate TDP constraint because it drops all LC tasks when the system switches to the HI mode, which means it has zero-percent QoS of LC tasks in the HI mode. On the other hand, methods of [6] and [15]

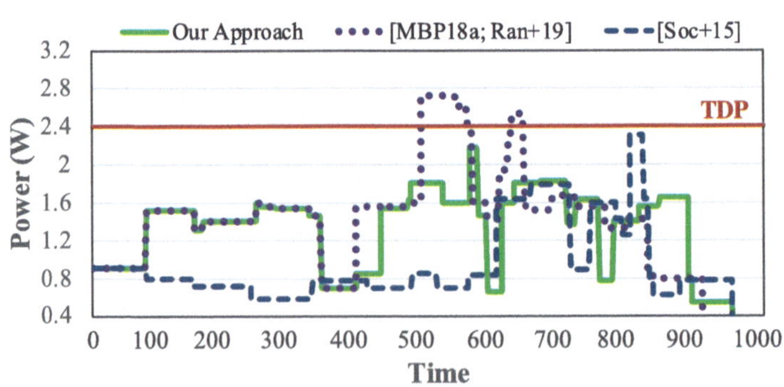

Fig. 6.6 Power trace of real-life application graph (CC) in different methods (proposed approach, [15] (Ran+19), [6] (MBP18a), and [14] (Soc+15)) under worst-case scenario

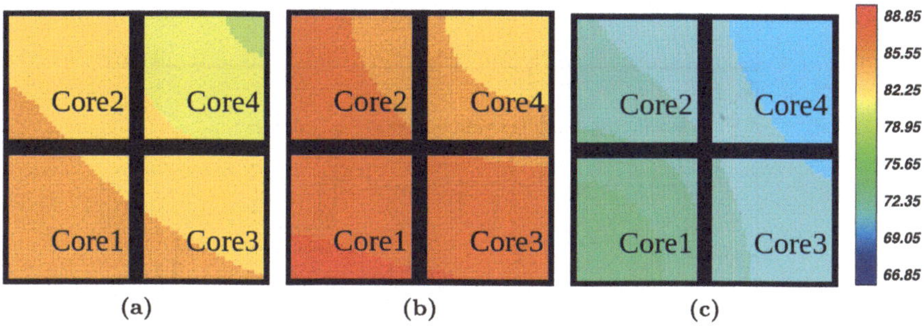

Fig. 6.7 Thermal profiles of real-life application graph (CC) in different methods (proposed Approach, [15] (Ran+19), [6] (MBP18a), and [14] (Soc+15)) under worst-case scenario. (**a**) Proposed method. (**b**) [6, 15]. (**c**) [14]

guarantee 90.91% of LC tasks' execution in the HI mode, but it frequently violates the TDP constraint (Fig. 6.6). For the CC application, our method endeavored to execute 81.82% of the LC tasks without violating TDP constraint. Figure 6.7 shows the steady-state temperature distribution of Socci [14], Medina [6], Ranjbar [15], and our proposed method using the HOTSPOT simulator for the CC application, corresponding to power profile of Fig. 6.6. Although the maximum temperature of [14] is lower than ours, it has zero-percent LC tasks' QoS because it does not execute any LC task in the HI mode. In addition, we map the tasks on the cores more uniformly than Medina's method, which prevents hotspots in our approach. The proposed approach could reduce 5 °C in maximum temperature compared to [6, 15]. If the system becomes larger in terms of the number of cores and tasks, the efficiency of our proposed approach in reducing the hotspots would be higher.

As a result, our method reduces the peak power and maximum temperature by up to 20.06% and 3.71% respectively, compared to the approach of [6, 15], while the QoS is degraded 9.09%. On the other hand, although our method increases the maximum temperature by 9.61%, compared to [14], we reduce the peak power consumption by 6.31% and improve the QoS by 81.82%.

Now, we show the power trace (Fig. 6.8) and heat map of the system (Fig. 6.9), when the Autoware application is running on the system with three cores. This application consists of 19 tasks; 10 HC tasks and 9 LC tasks. Similar to what is considered for the CC application, to focus on the behavior of systems in the HI mode, we assumed no fault occurred during the application's execution. According to the dependencies and timing of the tasks, all LC tasks could be scheduled before their deadlines in the proposed approach and methods of [6, 15]. As can be seen in these two figures, the proposed approach can manage the peak power consumption to be less than TDP constraint, compared to [6, 15] approaches. Note that the system peak power and maximum temperature under [14] approach is less in comparison to our approach due to executing fewer tasks (i.e., dropping LC tasks) when the system switches to the HI mode. As a result, the proposed method

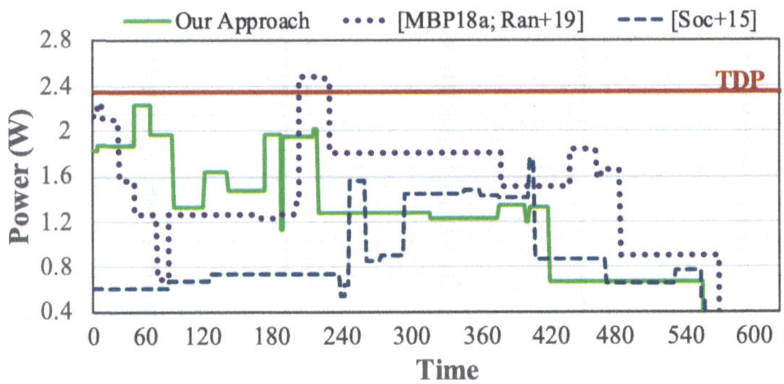

Fig. 6.8 Power trace of real-life application graph (Autoware) in different methods (proposed Approach, [15] (Ran+19), [6] (MBP18a), and [14] (Soc+15)) under worst-case scenario

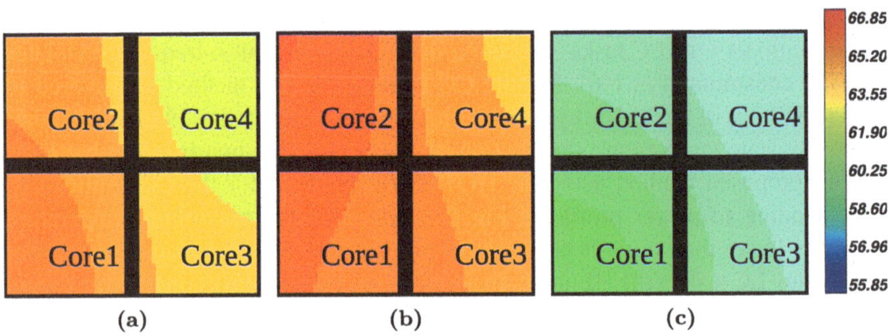

Fig. 6.9 Thermal profiles of real-life application graph (Autoware) in different methods (proposed Approach, [15] (Ran+19), [6] (MBP18a), and [14] (Soc+15)) under worst-case scenario. (**a**) Proposed method. (**b**) [6, 15]. (**c**) [14]

could reduce the peak power and maximum temperature by up to 9.9% and 3.2%, respectively, compared to the approach of [6, 15].

Now, we evaluate the power trace and thermal distribution of different methods of [6, 14, 15] and our proposed method for a random task set example in the worst-case scenario, in terms of execution times and power consumption. It should be noted that the scale of temperature for each method is different in Fig. 6.11. Since we have generated many task graphs with different values of parameters (n, d, c, and U/c), we choose one of the random task graphs with $d = 10\%$, $n = 50$, $c = 8$, and $U/c = 0.9$, which needs a high computational demand to show the results. Figure 6.10 shows the power traces, and Fig. 6.11 shows the thermal distribution of the methods. As can be seen in Fig. 6.10, the peak power consumption is violated sometimes in the method of [6, 15], and since the method of [14] drops all LC tasks in the HI mode, which is not desirable, the task set finishes its execution earlier and also has less peak power consumption. Besides, Fig. 6.11 depicts that the thermal distribution has not been managed in [6, 15], while our approach reduces the

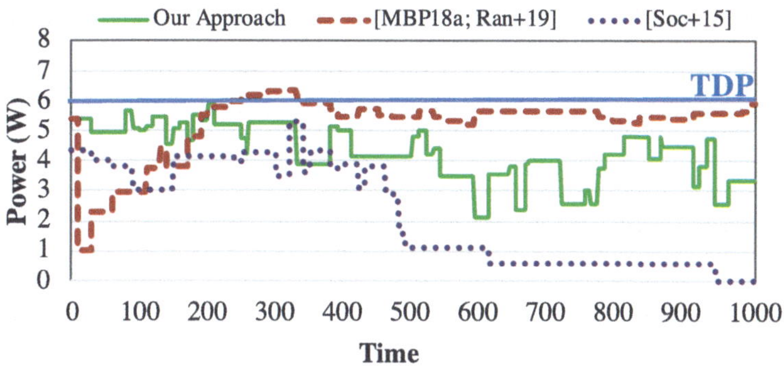

Fig. 6.10 Power trace of a random task graph in different methods (proposed Approach, [15] (Ran+19), [6] (MBP18a), and [14] (Soc+15)) under worst-case scenario

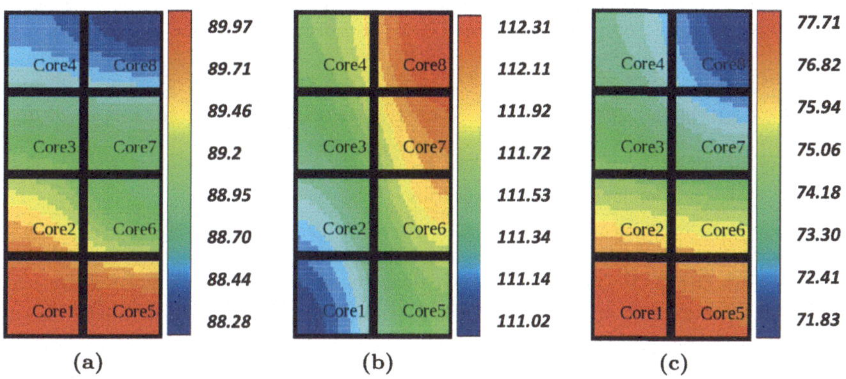

Fig. 6.11 Thermal profiles of a random task graph example in different methods (proposed approach, [15] (Ran+19), [6] (MBP18a), and [14] (Soc+15)) under worst-case scenario. (**a**) Proposed method. (**b**) [6, 15]. (**c**) [14]

hotspots and lowers the maximum temperature by 22.3 °C in this example. Although the maximum temperature of [14] is lower than ours, the LC tasks' QoS is zero since no LC tasks are executed in the HI mode.

6.4.5 *Analyzing the QoS of LC Tasks*

Now, we analyze the QoS for the proposed method in comparison with methods of [6, 14, 15] in Fig. 6.12. In our proposed method, there are many possible scenarios (each node of the tree is responsible for keeping the scheduling of the system in one scenario), and for each scenario, the LC tasks' QoS is different. Thus, we run 100 schedulable task sets on eight cores, and for each task set, ten different random situations of occurring faults and mode switching are considered. For the case when

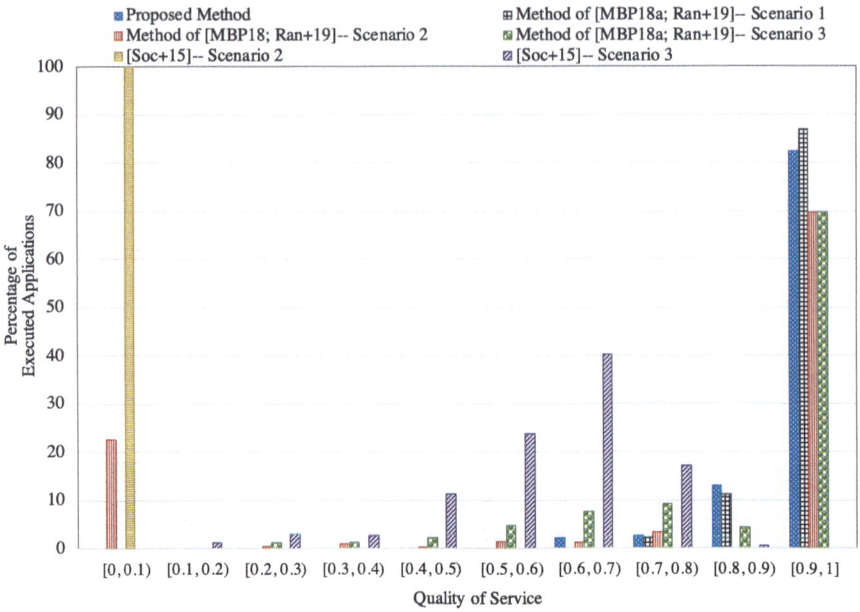

Fig. 6.12 Task sets' QoS under different scenarios

$U/c = [0.5, 0.75)$ or $U/c = [0, 0.5)$, there is more free slack before the task set deadline; therefore, in the case of fault occurrence and mode switching, fewer LC tasks are dropped. In this experiment, we consider the worst-case scenario of processor demands, $U/c = [0.75, 1]$, $n = 50$, and $d = 10\%$, and in the generated task sets, 20–50% of tasks are LC tasks. In addition, the number of fault occurrences in each scenario is randomly selected in the range of [0,4]. Figure 6.12 shows the LC tasks' QoS for all these 1000 scenarios. The QoS is the successfully executed LC tasks to all LC tasks. However, we use three different definitions of QoS to evaluate the methods of [6, 14, 15] more accurately, as follows:

- Scenario 1: The QoS refers to how many LC tasks are successfully executed before their deadlines with no TDP violation. If TDP is going to be violated, running LC tasks is only stopped to reduce the peak power consumption.
- Scenario 2: This scenario has the same definition as Scenario 1, with the difference that since the HC tasks are the most important, therefore without them, QoS of LC tasks is penalized by completely being zero. Thus, if TDP is violated and some HC tasks are running on cores, then the QoS = 0.
- Scenario 3: Since those methods have not been specifically designed for peak power management while meeting the real-time constraints of all HC tasks, we give the HC tasks higher weight and then consider the joint QoS, including both LC and HC tasks. Therefore, the HC tasks have a double weight in this scenario compared to LC tasks, in the case of dropping tasks due to the TDP violation.

As shown in Fig. 6.12, the QoS for methods of [6, 15] in Scenario 1 is higher than the QoS for our proposed method in total (according to the CDF line). However, in this scenario, the TDP constraint is violated several times due to the execution of HC tasks in parallel on cores, and there is no policy to manage the peak power. Moreover, for the second scenario in the methods of [6, 15], the figure shows that in 22.5% of task sets, the TDP constraint is violated while executing some HC tasks on cores. Besides, the QoS in the second scenario for [14] is zero, due to dropping all LC tasks in the HI mode. Now, if we investigate the methods of [6, 14, 15] in the third scenario, the QoS of [6, 15] is more than the QoS of [14] due to executing LC tasks in the HI mode. However, they have less QoS in this scenario compared to our proposed method. According to this figure, we can conclude that our proposed method is more efficient in improving QoS than other methods in different scenarios. In addition, the minimum and maximum QoS in our proposed method are 68.33% and 100%, while the power constraint is always met.

6.4.6 Peak Power Consumption and Maximum Temperature Analysis

In this section, we evaluate the system's peak power consumption and the chip's maximum temperature in our approach and methods of [6, 15] by varying different parameters, such as utilization bound (U/c), number of tasks (n), edge percentage (d), and the number of cores (c). Figure 6.13 shows the average normalized peak power consumption to the TDP constraint and maximum temperature in the worst case that the system switches to the HI mode after executing the first HC task and all tasks execute up to their higher WCET. Hence, in the worst-case scenario, [15] has the same power profile and thermal distribution as [6].

First, we analyze the peak power consumption by varying different parameters. The figure shows that our proposed approach can manage the peak power consumption to be less than the TDP constraint in all scenarios, while Medina's method violates the TDP constraint in almost all scenarios. In general, the impact of our approach is increased as the probability of using parallelism in the execution of tasks on cores is increased (a large number of cores (larger c) or less dependency between tasks (lower d)). Since the number of tasks and utilization is not changed by increasing c, the maximum power consumption by [6] is also reduced. However, since our proposed method endeavors to distribute the tasks on all cores to minimize hotspots and also minimize the instantaneous power consumption, our proposed approach in decreasing the peak power consumption is more efficient, compared to [6], while the c is increased. Besides, by reducing the dependency between tasks (d), although the system peak power consumption is increased and TDP is violated in [6], our approach always guarantees that the TDP constraint is not violated. In addition, although increasing the number of tasks or utilization increases the system's peak power consumption because the system does more computation, our approach guarantees that the TDP constraint will never be violated.

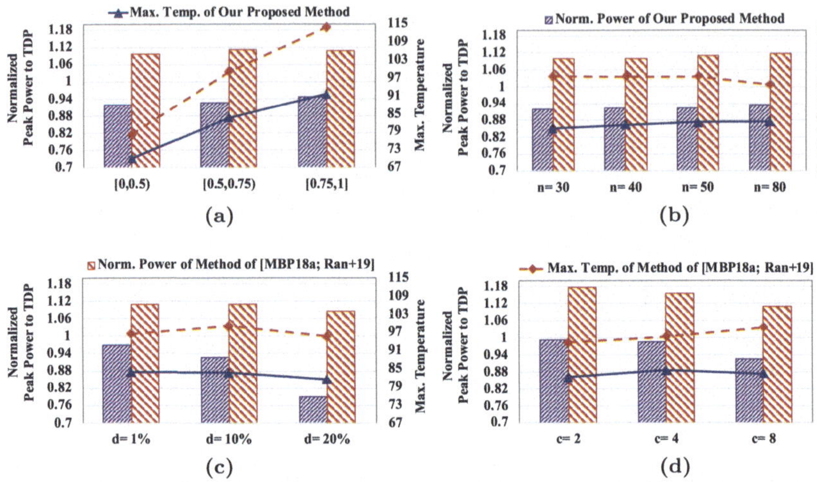

Fig. 6.13 Peak power consumption and maximum temperature in different methods (proposed approach, [15] (Ran+19), and [6] (MBP18a)). (**a**) Varying utilization. (**b**) Varying number of tasks. (**c**) Varying edge percentage. (**d**) Varying number of cores

From the perspective of maximum temperature analysis, increasing the system utilization, illustrated in Fig. 6.13a, while the number of tasks is fixed ($n = 50$) means that the task's execution time tends to be longer. Thus, the computation time of cores is increased, the managing of peak power constraint and busy/idle times of cores would be difficult, and consequently, the maximum temperature of the chip is increased in both methods. However, we can decrease the maximum temperature by up to 22.4 °C in comparison with [6, 15]. Besides, if we vary the number of tasks in Fig. 6.13b since the computation time of cores does not change, the chip's maximum temperature is relatively constant by increasing the number of tasks in both our proposed method and methods of [6, 15]. Additionally, by varying d, the maximum temperature reduces by increasing the dependency between tasks because the cores' computation time is constant, while the idle time of cores is increased. As shown in Fig. 6.13c, our proposed method can reduce the maximum temperature by 14.3 °C on average by varying edge percentage, compared to [6, 15].

Now, we investigate the system's maximum temperature in Fig. 6.13d by increasing the number of cores (c), while other parameters are constant. Hence, the normalized utilization (U/c is constant in this experiment, which means the utilization (U) is increased by increasing c. We have more parallelism to execute tasks by increasing c and add more free slack to let the cores be idle (having better thermal distribution). However, since each core's temperature is a function of its neighbor cores' temperature, it increases the chip's maximum temperature, while the system utilization is increased. Therefore, the results of our proposed method show that the maximum temperature is relatively constant by increasing c and can decrease it, 10.7 °C on average, compared to [6, 15]. Since the methods of [6, 15] do not

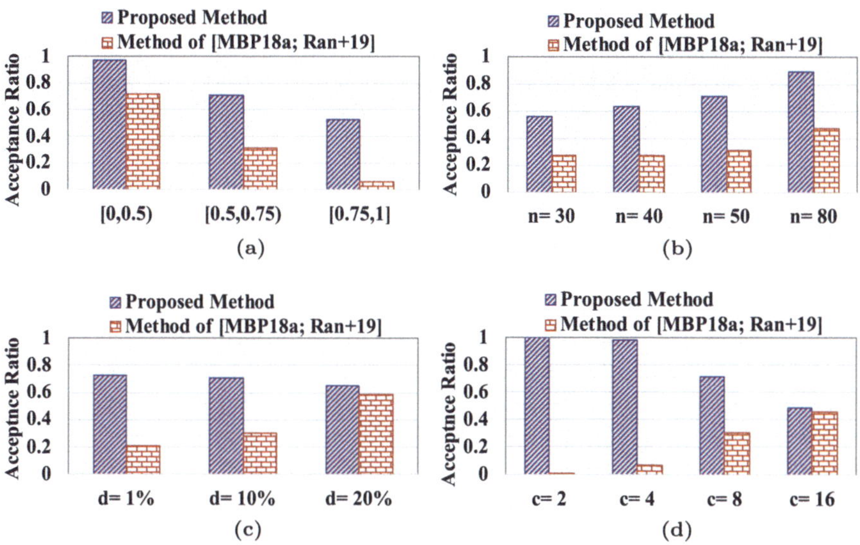

Fig. 6.14 Normalized acceptance ratio under different scenarios in different methods (proposed approach, [15] (Ran+19), and [6] (MBP18a)). (**a**) Varying utilization. (**b**) Varying Number of tasks. (**c**) Varying edge percentage. (**d**) Varying number of cores

consider the thermal distribution, the maximum temperature is generally increased by increasing c.

6.4.7 Effect of Varying Different Parameters on Acceptance Ratio

In this section, we illustrate the impact of different parameters, such as utilization bound (U/c), number of tasks (n), edge percentage (d), and the number of cores (c) on the task schedulability (acceptance ratio). Figure 6.14 represents the effect of each parameter, while the others are fixed, to analyze how the proposed method and the methods of [6, 15] react to each parameter. We run 1000 benchmarks for each scenario and report the average result. A task set is schedulable if the real-time and power constraints are met. In general, having more dependency between tasks, large system utilization, or more cores causes the system to have less acceptance ratio in our proposed method. We discuss the observation in detail.

From the perspective of utilization bound, we fix other parameters to see the effect of varying utilization in Fig. 6.14a. Increasing the utilization while the number of tasks is fixed ($n = 50$) means that the tasks' execution time tends to be longer. When the utilization is getting higher, the computation time of cores is increased. Therefore, fewer task sets can be scheduled before their deadlines even in the case of fault occurrence, and also, management of power constraint and busy/idle times of cores would be difficult. We can conclude that fewer task sets can be scheduled before their deadline, while the power constraint is not violated. This trend is also the same with [6, 15] with the difference that the TDP constraint is violated several times, which causes the task sets not to be schedulable.

Besides, Fig. 6.14b shows that the task schedulability is increased by increasing the number of tasks, with $d = 10\%$, U/c= [0.5,0.75), and $c = 8$. Since the system utilization is constant for all number of tasks, the execution time of tasks is reduced by increasing the number of tasks. Therefore, the tasks tend to finish their execution early and allow their successors to be released. Furthermore, the overhead of re-executing a task due to fault occurrence is much lower for small tasks. According to Fig. 6.14b, we conclude that our method can schedule 70% of task sets, on average, when only the number of tasks is varied. Besides, by increasing the number of tasks in the methods of [6, 15], the parallel task execution is increased, which causes more peak power consumption, and therefore, less task schedulability due to the TDP violation.

For the case of varying the edge percentage, when the dependency between the tasks is increased, while n, U/c, and c are constant ($n = 50$, $U/c = [0.5, 0.75)$, $c = 8$), the release time of tasks is increased, because tasks must wait for more predecessor tasks to finish their executions. Therefore, the idle time on cores increases, which causes more delays in the execution of tasks, and reduces the schedulability. Figure 6.14c shows that the highest schedulability (73%) is achieved in our method when $d = 1\%$. Figure 6.14d shows the effect of varying the number of cores in the system on task schedulability when other parameters are not changed ($n = 50$, $U/c = [0.5, 0.75)$, $d = 10\%$). By considering the fixed U/c, the utilization is increased by increasing the number of cores. Consequently, the execution time of tasks is increased because the number of tasks is fixed. As mentioned earlier, task schedulability is decreased by increasing the tasks' execution time. Therefore, as can be seen in Fig. 6.14d, the schedulability of applications with our proposed method decreases by increasing the number of cores, while the other parameters are fixed. Besides, in methods of [6, 15], the acceptance ratio increases by increasing the dependency between tasks and having more cores in the system. The reason is that based on their mapping and scheduling algorithm, by increasing the edge percentage and number of cores while fixing other parameters, tasks have less parallelism, and also fewer cores are selected to be active to execute the tasks, which causes the system to have less peak power consumption and therefore, a higher acceptance ratio. As a result, by increasing the edge percentage for more than 20%, our proposed method and method of [6, 15] have almost the same acceptance ratio. However, the mapping and scheduling algorithm of [6, 15] increases the overheating problem, which is not acceptable by most safety-critical systems. In addition, since methods of [6, 15] are not peak power-aware, when the number of cores is less, the TDP is violated in most of the task sets, and therefore, these task sets are not schedulable.

In the end, the acceptance ratio of our proposed method is 74.14% on average for all scenarios, while it is 31.1% in the methods of [6, 15].

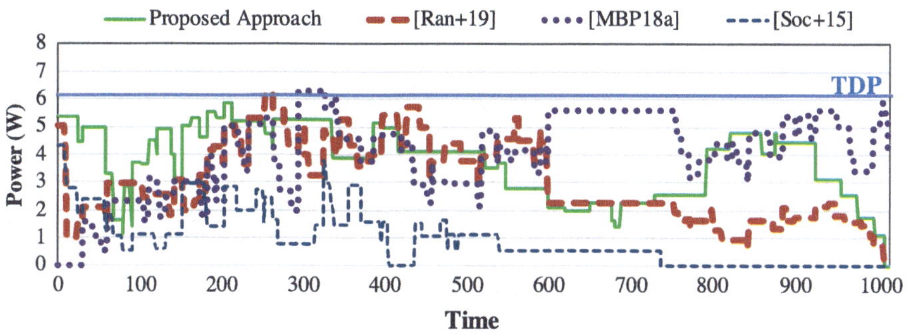

Fig. 6.15 System power trace of different methods (proposed approach, [15] (Ran+19), [6] (MBP18a), and [14] (Soc+15)) for a random task graph at run-time

6.4.8 Investigating Different Approaches at Run-Time

Now, we evaluate the system behavior at run-time in terms of peak power consumption for our proposed approach and the method proposed in [6, 14, 15]. We in [15] have presented a run-time method to reclaim the available slacks and reduce the *V-f* levels of cores to decrease the system peak power. Here, analogous to [15], the actual execution time of tasks follows the normal distribution with the mean and standard deviation of $\frac{3\times WCET}{4}$ and $\frac{WCET}{12}$, respectively. Figure 6.15 depicts the run-time power trace of methods for a random task graph with $d = 10\%$, $n = 50$, $c = 8$, and $U/c = 0.9$. The system switches to the HI mode by forcing a randomly selected HC task to execute beyond its lowest WCET for both methods. As shown in this figure, the system peak power in the proposed approach is less than the TDP constraint at run-time, while the method of [6] has violated the TDP for a period of time. Although the system peak power of [15] may be less than the TDP constraint for some applications like the used task graph and their method consumes less energy in the system, there is no guarantee for the peak power to be less than the power constraint. Due to the use of DVFS technique in [15] and decreasing the *V-f* levels at run-time, the system consumes less energy in comparison with our proposed method. For the example of Fig. 6.15, the method of [15] saves $0.74102J$ in the system compared to our proposed approach. However, the DVFS technique degrades the reliability and increases the fault rate. The fault rate depends on the system's voltage level, and also, the application's reliability depends on the voltage level and tasks' WCET, which is increased by reducing the frequency level [10]. As an example, for this task graph, by considering the fault rate $f = 10^{-4}$ [18], the reliability of tasks has been decreased 0.17% and 2.08%, on average and worst case in comparison with our proposed approach. In addition, the number of nines for the system's reliability ($-Log_{10}^{1-Rel}$) has been degraded from 8 to 6, which may not be desirable for most safety-critical applications [19].

From the perspective of the system's reliability and fault tolerance in our proposed method, we run the 1000 task graph applications for different normalized

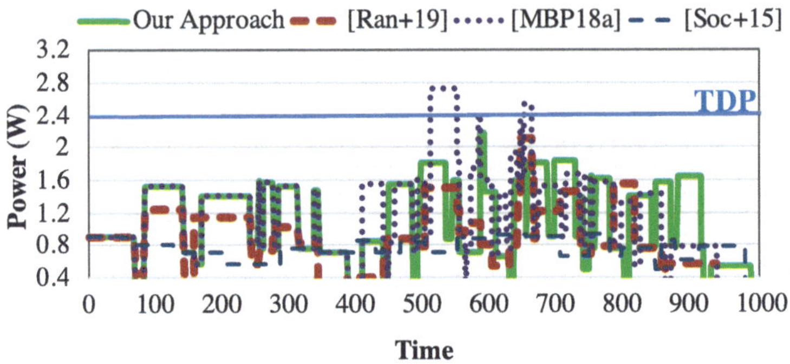

Fig. 6.16 Run-time power trace of real-life task graph in different methods (proposed approach, [15] (Ran+19), [6] (MBP18a), and [14] (Soc+15))

utilization bound ($U/c = [0.5, 0.75)$, and [0.75,1]) and compare to the system's reliability in [15]. In our proposed method, $-Log_{10}^{1-Rel}$ for the normalized utilization equal to [0.5,0.75) and [0.75,1] is 8.56 and 7.67 on average, respectively, while for [15], it is 4.99 and 4.60, respectively. As a result, based on the required reliability for safety-critical systems, the method of [15], which decreases the *V-f* levels, has severely damaged the system's safety, which is not desirable. The reliability of the proposed approach is high for different utilization. Therefore, our method can be applied to any application with varying bounds of utilization while satisfying peak power management, fault tolerance, and high reliability in different system operational modes.

Since we have used a real-life task graph (CC) to evaluate our proposed method, here we show the run-time system behavior for the real task graph in Fig. 6.16. As shown in this figure, the approach of [6] still violates the power constraint (TDP), and since the approach of [15] has applied the DVFS technique, the peak power consumption has been reduced. Hence, the method of [15] uses the same task mapping and scheduling algorithm of [6] at design-time and reduces the peak power by reclaiming the dynamic slack times at run-time (the difference between the WCET and actual execution time) and decreases the operating voltage and frequency level of cores while executing the tasks. However, the approach of [15] has degraded reliability due to the use of DVFS technique, which is not desirable for safety-critical applications. In addition, our proposed method could manage the peak power consumption in this real task graph, and also, since the method of [14] has dropped all LC tasks in the HI mode, the peak power consumption is also less than the TDP constraint.

6.5 Conclusions

This chapter has proposed an approach to schedule MC tasks in fault-tolerant systems in different operational modes to manage peak power by considering a thermal management policy. At run-time, depending on the fault-occurrence possibilities and criticality mode changes, the system faces different scenarios. We proposed an approach that develops a tree of schedules at design-time. Each node of the tree represents a scenario and contains scheduling of tasks in which all HC tasks and as many as possible LC tasks can be executed without violating the TDP. At run-time, a low overhead online scheduler selects the proper node to map and schedule tasks. The results show that the proposed technique can schedule 74.14% of task sets on average and significantly reduce peak power consumption (by guaranteeing the TDP constraint) in the worst-case scenario compared to the existing methods. Besides, this approach can also extend to multiple criticality levels. However, we first need to know the importance and functions in different criticality levels and how they can be dropped in the higher criticality modes without impacting system functionality. Then, based on it, propose an efficient task mapping and scheduling algorithm in order to manage real-timeliness, power, and maximum temperature while improving the QoS of tasks with different criticality levels.

Although this approach guarantees the real-time constraints in the worst-case scenario of task execution time at run-time, most task execution times are significantly shorter than their WCET. Indeed, this worst-case scenario rarely happens. Due to the early finish of the task's execution, the generated dynamic slack could be used for better power-aware MC system design in multi-core platforms. In the next chapter, we propose an online method that adapts to task execution time dynamism and employs the accumulated dynamic slack to reduce the peak power consumption and maximum temperature.

References

1. Waqaas Munawar et al. "Peak Power Management for scheduling real-time tasks on heterogeneous many-core systems". In: *Proc. of the International Conference on Parallel and Distributed Systems (ICPADS)*. 2014, pp. 200–209.
2. Sanjoy Baruah. "The federated scheduling of systems of mixed-criticality sporadic DAG tasks". In: *Proc. of IEEE Real-Time Systems Symposium (RTSS)*. 2016, pp. 227–236.
3. Micaiah Chisholm et al. "Supporting Mode Changes While Providing Hard-ware Isolation in Mixed-Criticality Multicore Systems". In: *Proc. of Real-Time Networks and Systems (RTNS)*. 2017, pp. 58–67.
4. Pengcheng Huang et al. "Energy efficient dvfs scheduling for mixed-criticality systems". In: *Proc. on Embedded Software (EMSOFT)*. 2014, pp. 1–10.
5. Behnaz Ranjbar et al. "Power-Aware Runtime Scheduler for Mixed-Criticality Systems on Multicore Platform". In: *IEEE Transactions on Computer-Aided Design of Integrated Circuits and Systems (TCAD)* 40.10 (2021), pp. 2009–2023. https://doi.org/10.1109/TCAD.2020.3033374.

6. R. Medina, E. Borde, and L. Pautet. “Availability enhancement and analysis for mixed-criticality systems on multi-core”. In: *Proc. on Design, Automation & Test in Europe Conference & Exhibition (DATE)*. 2018, pp. 1271–1276.
7. Viacheslav Izosimov, Petru Eles, and Zebo Peng. “Value-based scheduling of distributed fault-tolerant real-time systems with soft and hard timing constraints”. In: *Proc. of the IEEE Workshop on Embedded systems for real-time multimedia (ESTIMedia)*. 2010, pp. 31–40.
8. V. Izosimov et al. “Scheduling of Fault-Tolerant Embedded Systems with Soft and Hard Timing Constraints”. In: *Proc. on Design, Automation & Test in Europe Conference & Exhibition (DATE)*. 2008, pp. 915–920.
9. Y. Zhang and K. Chakrabarty. “A unified approach for fault tolerance and dynamic power management in fixed-priority real-time embedded systems”. In: *IEEE Transactions on Computer-Aided Design of Integrated Circuits and Systems (TCAD)* 25.1 (2006), pp. 111–125. https://doi.org/10.1109/TCAD.2005.852657.
10. M. Salehi et al. “Two-State Checkpointing for Energy-Efficient Fault Tolerance in Hard Real-Time Systems”. In: *IEEE Transactions on Very Large Scale Integration (VLSI) Systems* 24.7 (2016), pp. 2426–2437.
11. Wei Huang et al. “HotSpot: A compact thermal modeling methodology for early-stage VLSI design”. In: *IEEE Transactions on Very Large Scale Integration (VLSI) Systems* 14.5 (2006), pp. 501–513.
12. Y. Gong, J. J. Yoo, and S. W. Chung. “Thermal Modeling and Validation of a Real-World Mobile AP”. In: *IEEE Design & Test* 35.1 (2018), pp. 55–62.
13. Yongpan Liu et al. “Thermal vs energy optimization for dvfs-enabled processors in embedded systems”. In: *Proc. of International Symposium on Quality Electronic Design (ISQED)*. 2007, pp. 204–209.
14. Dario Socci et al. “Multiprocessor scheduling of precedence-constrained mixed-critical jobs”. In: *Proc. of International Symposium on Real-Time Distributed Computing (ISORC)*. 2015, pp. 198–207.
15. B. Ranjbar et al. “Online Peak Power and Maximum Temperature Management in Multi-core Mixed-Criticality Embedded Systems”. In: *Proc. of Euromicro Conference on Digital System Design (DSD)*. 2019, pp. 546–553.
16. Israel Koren and C Mani Krishna. *Fault-tolerant systems*. Morgan Kaufmann,2020.
17. AutowareAuto Project. https://gitlab.com/autowarefoundation/autoware.auto/AutowareAuto.git. Accessed: May 2021.
18. Mohammad Salehi, Alireza Ejlali, and Bashir M Al-Hashimi. “Two-phase low-energy N-modular redundancy for hard real-time multi-core systems”. In: *IEEE Transactions on Parallel and Distributed Systems (TPDS)* 27.5 (2016), pp. 1497–1510.
19. B. Ranjbar et al. “FANTOM: Fault Tolerant Task-Drop Aware Scheduling for Mixed-Criticality Systems”. In: *IEEE Access* 8 (2020), pp. 187232–187248. https://doi.org/10.1109/ACCESS.2020.3031039.

Chapter 7
QoS- And Power-Aware Run-Time Scheduler for Multi-core Mixed-Criticality Systems

In this chapter, we target peak power consumption and maximum temperature issues in MC systems with dependent tasks at run-time and evaluate the algorithm on a real multi-core platform. Although there are works that manage or minimize the power consumption of MC systems, they have not considered the instantaneous peak power consumption in both HI mode and LO mode, and their algorithms have often been limited to simulation. To solve the problem, we exploit dynamic slacks, the slack between tasks' AET and their WCET, along with DVFS at run-time, while the MC tasks' deadlines are guaranteed. Our approach has two phases: (1) at design-time, the tasks are scheduled on each core based on the EDF algorithm, and the resulting schedule is stored to be used as a static scheduling table. This is performed for both LO mode and HI mode. In this case, the number of LC tasks that have to be dropped in the HI mode is minimized to improve the overall QoS of the system. (2) At run-time, we examine multiple tasks in the future (look-ahead) to select the most appropriate one to assign the currently available dynamic slack. The selection is based on the impact of the tasks on the peak power and temperature of the system, which is quantified by a weighted multi-objective cost function. Therefore, the speed of the core that runs the task can be decreased accordingly using per-cluster DVFS. Additionally, besides exploiting the dynamic slacks, we propose a task re-mapping technique at run-time to improve the system temperature profile further. However, the online scheduler timing overheads for selecting an appropriate task and checking the re-mapping technique to choose a proper core are crucial for the MC systems and may cause deadline violations. Furthermore, the timing overhead of changing *V-f* levels in using the DVFS technique is critical in run-time task scheduling. Therefore, we analyze and evaluate the effect of these overheads on the schedule of MC tasks in real multi-core platforms. We study that these overheads cannot be neglected due to their impact on meeting MC tasks' deadlines. Besides, we optimize the run-time scheduler to minimize the timing overhead. In summary, the main contributions of this chapter are the following:

B. Ranjbar et al., *Quality-of-Service Aware Design and Management of Embedded Mixed-Criticality Systems*, https://doi.org/10.1007/978-3-031-38960-3_7

- An online peak power and maximum temperature management of MC systems in heterogeneous multi-core platforms while respecting deadline requirements of tasks in both LO mode and HI mode.
- A multi-task look-ahead approach to make sure that dynamic slacks are assigned to the tasks that lead to more peak power reduction and maximum temperature reduction.
- An online task re-mapping technique that exploits dynamic slacks to re-map the tasks to other cores within a cluster in order to lower the system temperature.
- Studying the online scheduler and DVFS governor in terms of timing overhead to provide the deadline guarantee of MC tasks during run-time phase.
- By measuring on a real platform we observe that while the latency of the scheduler is minimal (less than 10 μs on average), the latency of the DVFS switching is 5.313 ms on average and, thus, cannot be neglected.

In the following, the research questions, objectives, and motivational example are presented in Sect. 7.1. Then in Sects. 7.2 and 7.3, we present the design-time approach and propose our proposed method and algorithm in detail, respectively. The analysis and optimization of the run-time scheduler are then studied in Sect. 7.4. Finally, we analyze the experiments and conclude the chapter in Sects. 7.5 and 7.6, respectively.

7.1 Research Questions, Objectives, and Motivational Example

The crucial research questions that are addressed in this chapter are as follows:

1. How to select the most appropriate tasks to assign the dynamic slack to, for managing the peak power consumption?
2. Whether it is possible to re-map the tasks to other cores for better thermal control, and if yes, where and when should the tasks be re-mapped to?
3. Which timing overheads during run-time have an impact on task scheduling and deadline misses?
4. How these run-time timing overheads can be managed to not affect tasks' deadlines?

To clarify the problem and provide some insight into how a run-time scheduler can manage peak power consumption, a motivational example is given. Figure 7.1a shows a precedence constraint MC task graph with eight tasks mapped on two cores, and tasks' information such as WCETs and peak power consumption. Although each task in the graph can have a local deadline (the reader can refer to Sect. 2.1.1.1 for more detail about deadline definition in the task graph model), the whole task graph has the deadline of $D = 200$ ms. In addition, in order to simulate the variation in the actual run-time, the actual execution time values are selected from a uniform distribution of $[\frac{2}{3}.WCET, WCET]$, analogous to [1–3], which have used uniform

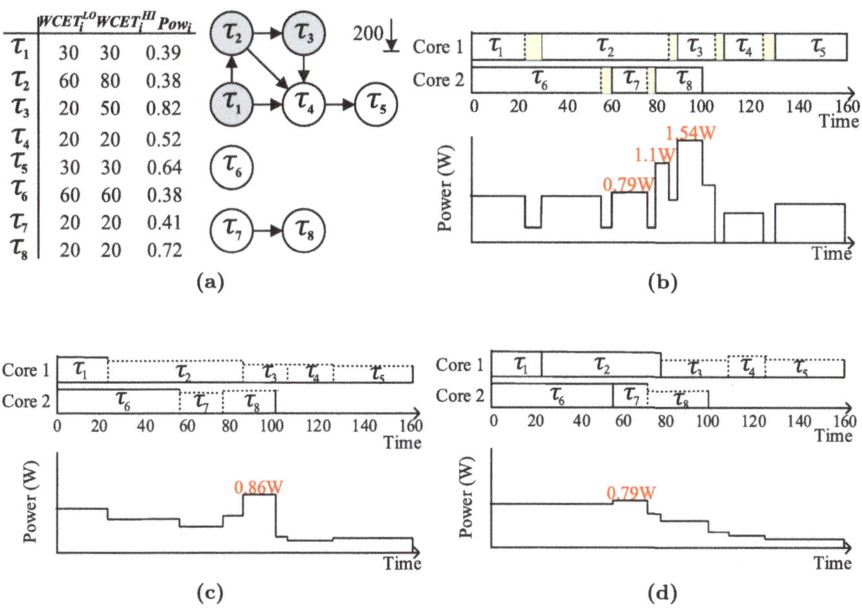

Fig. 7.1 A motivational example for a real-life application in different scenarios. (**a**) A task graph. (**b**) System power trace at run-time without using DVFS. (**c**) System power trace by using DVFS and considering one task look-ahead. (**d**) System power trace by using DVFS and considering two tasks look-ahead

distribution. We obtain the task mapping and scheduling table using the algorithm presented in [4]. In this example, we suppose that the system is only in the LO mode for simplicity of presentation. Besides, we assume that the tasks consume their maximum power continuously during their executions. Figure 7.1b shows the task schedule and the system power trace at run-time. In the worst-case scenario, the system's peak power may be high and may lead to thermal hotspots and instability, which has not been investigated in recent studies in MC systems that have different criticality modes. As shown in Fig. 7.1b, since the tasks may finish earlier than their WCET, for example, τ_1 at 22 ms while its WCET = 30 ms, the incurred slack can be exploited and assigned to the following tasks to reduce the peak power consumption. In Fig. 7.1c, these dynamic slacks are used for the immediately ready tasks (one task look-ahead) to decrease the speed of its corresponding core. Some power reduction can be observed. However, in some cases, the immediate task that follows may consume much less power than the other tasks after that. Therefore, it is better to reserve that slack for the task after that if it is possible. As shown in Fig. 7.1d, if we select the task by looking two tasks ahead, more peak power reduction can be achieved as compared to Fig. 7.1c. Therefore, by comparing the maximum power consumption of two scenarios by using the DVFS) technique by looking two tasks ahead (Fig. 7.1d) and without using DVFS (Fig. 7.1b), we have 48.7% reduction in peak power consumption. In addition, we have a 20.12% reduction in energy consumption.

Equation (7.1) presents the goal of our proposed method, minimizing the peak power and the maximum temperature of individual cores during run-time:

$$Minimize(Pow_j, (Temp_{max})_j)|_{(j \in Cores)} \tag{7.1}$$

As shown in the motivational example, DVFS is one of the techniques that we use to manage the metrics (peak power and maximum temperature). Reducing the *V-f* level of a core while executing a task increases the execution time of the task and may cause deadline violation. In addition, the latency of changing *V-f* level or run-time scheduling may cause deadline violation. Equation (7.2) represents that the sum of the execution time of each task i on the core j at the *V-f* level l and timing overheads of the run-time scheduler ($TO_{Sch.}$) and changing *V-f* level (TO_{Vf}) must not exceed the task deadline in each criticality mode:

$$TO_{Sch.} + TO_{Vf} + \frac{WCET_i}{f_{jl}} \leq d_i \rightarrow \begin{cases} WCET_i = WCET_i^{LO} & \text{if } mode = LO \\ WCET_i = WCET_i^{HI} & \text{if } mode = HI \end{cases} \tag{7.2}$$

The proposed approach consists of design-time and run-time phases. It is worth noting that the proposed method takes advantage of the run-time phase to manage the peak power and temperature; hence, it is not possible to use any optimization method such as Integer Linear Programming (ILP) due to its long execution time. Thus, we develop a heuristic-based method. Figure 7.2 shows the flow of our proposed approach, along with the hardware platform. The hardware platform is used in design-time phase for tasks' power profiling and in run-time phase for the execution of tasks on cores. Now, we explain our approach comprising of the design-time and run-time phases, in detail.

7.2 Design-Time Approach

The input to the algorithm is a precedence-constrained task set and the multi-core system description, as shown in Fig. 7.2. The power required by the tasks can be obtained by running the benchmarks on a real platform, which is discussed in detail in Sect. 7.5. It should be noted that handling an unknown application during run-time is beyond the scope of this work. Since we target embedded applications, normally, the designer knows the system's tasks and their parameters at design-time. Therefore, by using the parameters of MC tasks such as WCETs, two tables of static task mapping and scheduling for LO mode and HI mode are created as shown in the design-time phase of Fig. 7.2. EDF algorithm is used to calculate the schedule of the tasks in each of the two modes statically based on the WCETs of LC and High-Criticality (HC) tasks, using the algorithm presented in [4]. In the LO mode,

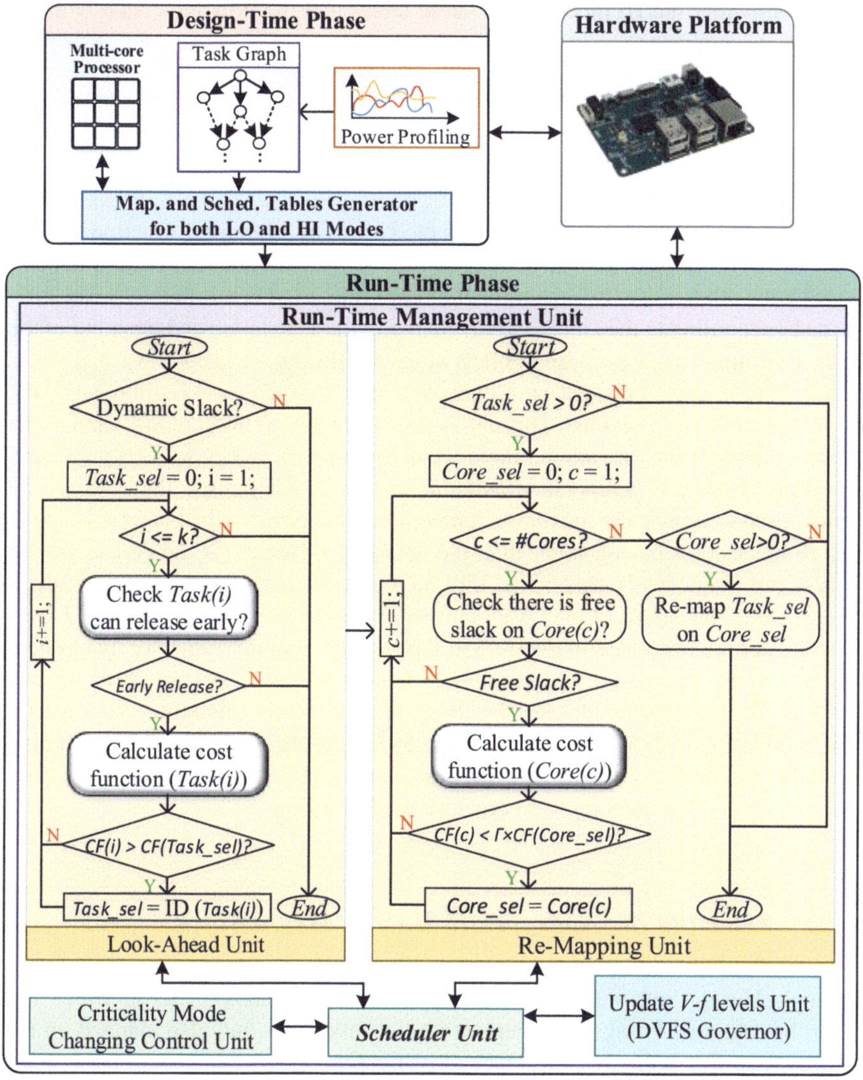

Fig. 7.2 Overview of our proposed approach

all tasks are scheduled with equal priority; in the HI mode, HC tasks are scheduled with a higher priority. These static schedules in the respective modes are then used to execute all tasks at run-time. This enforces a strict ordering in the execution of the tasks and guarantees that all deadlines are met according to the design-time analysis in both modes. It should be noted that since the WCETs of HC tasks are higher in the HI mode, not all LC tasks may be schedulable in the HI mode. In order to maximize the overall QoS, the algorithm tries to drop as few LC tasks as possible when computing the HI mode table. These tables and the info associated with the tasks are used during the run-time phase by our algorithm to manage the system.

7.3 Run-Time Mixed-Criticality Scheduler

The run-time phase of our proposed method consists of several function control units, as shown in Fig. 7.2. The Scheduler Unit is the main unit that is communicating with the other units. Two main functions are supported in this unit: (1) executing the tasks according to the tables and (2) changing the scheduling and mapping of the tasks according to our proposed policy which we discuss in Sects. 7.3.1 and 7.3.2. When there is any free slack on a core, or a task finishes its execution early, the Look-Ahead Unit is executed. This unit is used to choose a subset of tasks and select the most appropriate one among them. If an appropriate task is selected in a core, according to the core temperature and temperature of other cores, the Re-mapping Unit is used to reduce the maximum temperature and decide whether to re-map the task to other cores or not. After that, the obtained *V-f* level for the core is stored. This stored frequency is used by the DVFS Governor Unit when the task is ready to be executed. The details of the DVFS Governor Unit to select the optimum *V-f* level for a cluster are discussed in Sect. 7.3.4. Due to MC systems' behavior, the system switches to the HI mode if the execution of at least one HC task exceeds its defined $WCET^{LO}$. It should be checked by the Criticality Mode Changing Control Unit presented in Fig. 7.2. In this case, the system changes its task scheduling according to the HI scheduling table which is generated at design-time. The details for Look-Ahead Unit and Re-mapping Unit are described as follows.

7.3.1 *Selecting the Appropriate Task to Assign Slack*

In Look-Ahead Unit, we consider an approach named look-ahead in which our algorithm chooses k tasks after generating dynamic slack and also mapped on the same core in which the dynamic slack (S_{dyn}) is generated.[1] For each of the k tasks, a cost function is computed as defined by Eq. (7.3):

$$CF_i = \alpha \times E_i + \beta \times Pow_i \tag{7.3}$$

In this function, Pow_i and E_i are the maximum instantaneous power and maximum energy, respectively, that a task consumes to execute. In addition, α and β are in the range of [0,1]. Besides, energy reduction leads to a decrease in chip temperature [5]. Note that, if we consider $\langle \alpha, \beta \rangle = \langle 0, 1 \rangle$, the cost function only

[1] Finding the optimum value for k is discussed in Sect. 7.5.

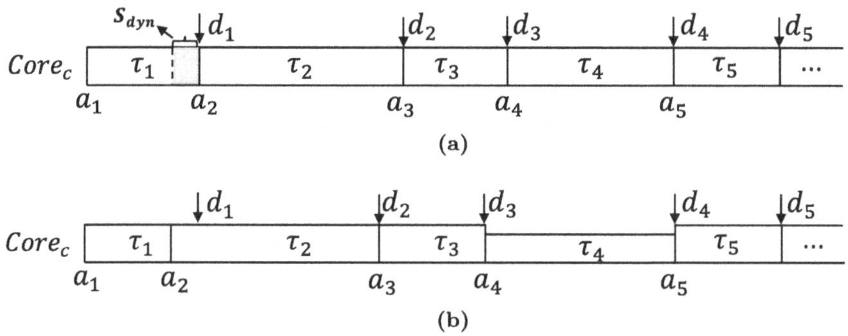

Fig. 7.3 An example of look-ahead policy. (**a**) Before assigning slack. (**b**) After assigning slack

considers the power of a task, and not its energy. Hence, the task with the largest peak power consumption is chosen to be executed at reduced core speed, in order to reduce the peak power consumption. If we have $\langle \alpha, \ \beta \rangle = \langle 1, 0 \rangle$, only energy is considered in the cost function. Hence, the task with the largest energy consumption is chosen to be executed at reduced core speed, thereby reducing the maximum energy consumption After selecting the task, the maximum power consumption and its WCET ($WCET_i^{LO}$ or $WCET_i^{HI}$) are changed based on the size of generated slack time and the *V-f* level. As a result, the start time and the deadline of tasks that are executed between the generated dynamic slack and selected task are *shifted* left based on the amount of slack to let the chosen task run with less speed, for example, tasks τ_2 and τ_3 illustrated in Fig. 7.3.

Furthermore, Eq. (7.3) is applied to a set of tasks that can start their executions earlier. A task (τ_i) can start early if it is released before $a_i - S_{dyn}$, where a_i is the start time of τ_i. As mentioned, a task can be released when all its predecessors finish their execution. Therefore, we just check $\tau_{ri} \leq \ a_i - S_{dyn}$, where τ_{ri} is the release time of τ_i. Consider the selected task τ_i with the start time a_i and deadline d_i that $a_i + WCET_i \leq \ d_i$. Assuming that we have the amount of slack, S_{dyn} generated by τ_j, during run-time. To utilize this slack time for the appropriate task τ_i, in general, the scheduler finds the minimum acceptable frequency based on $f_i = max(f_{min}, \frac{WCET_i}{WCET_i + S_{dyn}} . f_{max})$. This ensures that only the start time of the task is earlier by S_{dyn} and the deadline is kept unchanged, for example, τ_4 shown in Fig. 7.3. Hence, $a_i - S_{dyn} + \frac{WCET_i}{f_i / f_{max}} \leq a_i + WCET_i \leq d_i$. However, as mentioned at the beginning of this section, selecting the proper task and the core and changing the *V-f* level have overheads.[2] If we ignore them while selecting the optimum frequency, it may cause a deadline violation. Therefore, S_{dyn} is reduced by $TO_{Sch.}$ and TO_{Vf}. After selecting the optimum frequency, the start time of the appropriate task τ_i (a_i) is updated for the static schedule.

[2] We discuss in Sect. 7.5 how these timing overheads ($TO_{Sch.}$ and TO_{Vf}) are measured.

Now, we show the proof of the optimal solution of peak power minimization in individual cores, when $\langle\alpha,\ \beta\rangle = \langle 0, 1\rangle$ in Eq. (7.3) to select the task solely based on its peak power consumption. Let us assume that the task τ_i finishes its execution at time f_i, ahead of its deadline d_i, and a dynamic slack ($S_{dyn} = d_i - f_i$) is generated. The algorithm looks k tasks after generated slack to select the appropriate task and use the generated slack to reduce its *V-f* level and, consequently, decrease its power consumption. Without loss of generalization, assume that task τ_{i+l} consumes the highest peak power in the core within the k tasks looking ahead, presented in Eq. (7.4). This equation can be rewritten as Eq. (7.5), in which $Max(\tau_{i+1}^{pow}, \ldots, \tau_{i+l-1}^{pow}, \tau_{i+l+1}^{pow}, \ldots, \tau_{i+k}^{pow}) < \tau_{i+l}^{pow}$:

$$Pow_{core}^{max}|_{[d_i, d_{i+k}]} = Max(\tau_{i+j}^{pow})|_{j=1:k} = \tau_{i+l}^{pow},\ 1 \leq l \leq k \tag{7.4}$$

$$Pow_{core}^{max}|_{[d_i, d_{i+k}]} = Max(Max(\tau_{i+1}^{pow}, \ldots, \tau_{i+l-1}^{pow}, \tau_{i+l+1}^{pow}, \ldots, \tau_{i+k}^{pow}), \tau_{i+l}^{pow}) \tag{7.5}$$

If $\tau_{i+l}^{pow'}$ is the maximum power consumption of task τ_{i+l} after reclaiming the slack and reducing the *V-f* level, then $\tau_{i+l}^{pow'} < \tau_{i+l}^{pow}$; therefore, the core's maximum power consumption can be written as follows, which is less than τ_{i+l}^{pow}:

$$Pow_{core}^{max}|_{[d_i, d_{i+k}]} = Max(Max(\tau_{i+1}^{pow}, \ldots, \tau_{i+l-1}^{pow}, \tau_{i+l+1}^{pow}, \ldots, \tau_{i+k}^{pow}), \tau_{i+l}^{pow'}) \tag{7.6}$$

If we select one of the other tasks between $\{\tau_{i+1}, \ldots, \tau_{i+l-1}, \tau_{i+l+1}, \ldots, \tau_{i+k}\}$, and reduce its *V-f* level and consequently, its power consumption, then the peak power of the core is still limited by τ_{i+l}^{pow} according to Eq. (7.5). Hence, this power consumption is more than the optimum power consumption obtained by Eq. (7.6): τ_{i+l} is, therefore, the optimum task to which the slack should be assigned (given the constraint that all the slack is assigned to *one* of the following k tasks). We conclude that whenever a dynamic slack is generated, the proposed approach for selecting the appropriate task provides the optimum solution to minimize the peak power consumption of individual cores in the run-time phase.

Figure 7.4 shows a part of a static schedule of tasks on a core. Based on the peak power consumption of tasks in Fig. 7.4a, task τ_{i+4} is the appropriate task which consumes the highest peak power in the core in the time interval $[d_i, d_{i+4}]$. Therefore, assigning the slack to this task will lower the peak power to below 4W (if we have a dynamic slack ($S_d = d_i - f_i = 5$), then $Power_{core}^{max} = 3W$ after slack assignment, shown in Fig. 7.4b). If we assign the generated slack to one of the other tasks (e.g., τ_{i+1}) instead, then the peak power of the core is still limited by τ_{i+4}, i.e., 4W, as can be seen in Fig. 7.4c.

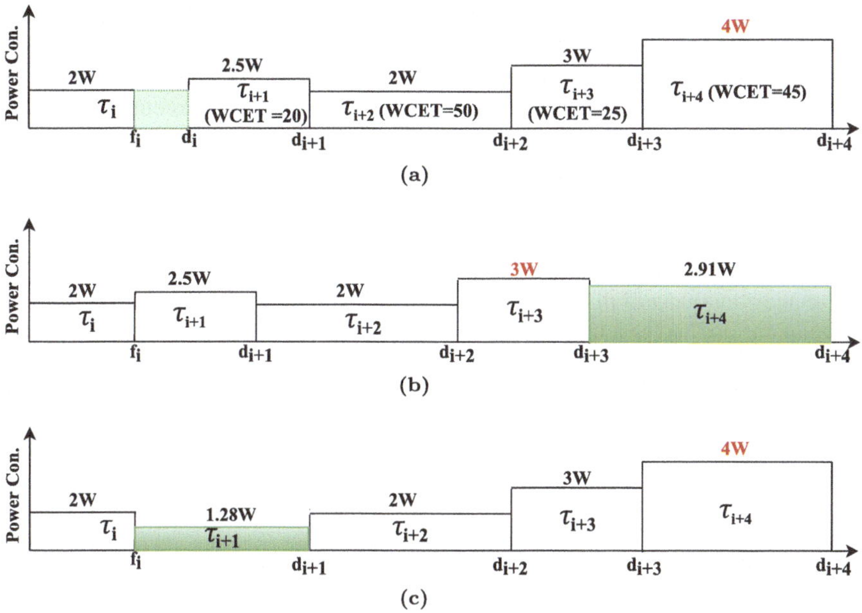

Fig. 7.4 Power trace of a core. (**a**) Power trace of the core. (**b**) Power trace of the core after assigning the slack to the task τ_{i+4}. (**c**) Power trace of the core after assigning the slack to the task τ_{i+1}

7.3.2 *Re-mapping Technique*

In order to manage the maximum temperature of the system and have better thermal control, it is possible to re-map the selected task to the other cores without changing its deadline. Therefore, to decide about re-mapping the task and selecting the appropriate core to re-map, we use the cost function in Eq. (7.7):

$$CF_c = \Gamma \times \sum_{t=1}^{t_f} E_c(t) \tag{7.7}$$

In this cost function, instead of using actual core temperature, we predict their temperature according to the accumulated energy. Based on our observation, a core tends to have a lower temperature when its accumulated energy is less than the other cores.[3] However, the difference between the accumulated energy of the base core and the selected core should be large enough. Therefore, we define a coefficient (Γ), which is equal to 0.9 in our experiments. In this equation, t_f is the time when any

[3] We show the observation about the relevance between core temperature and its accumulated energy consumption in Sect. 7.5.

particular task is finished. Besides, in order to not affect the tasks' deadline mapped on other cores, cores are examined for re-mapping that have free slack at the same period to execute the appropriate task. Since we consider the clustered multi-core platform (ODROID XU3) for our experiment, each application's execution time and power consumption will be different when running on different clusters. Hence, we use the re-mapping technique within each cluster. The reason is that although re-mapping from a little core to a big core reduces a task's execution time, it causes the system's peak power consumption to increase, which is not acceptable based on our targets. Therefore, to not change the system peak power consumption, we use the re-mapping technique within each cluster. It should be noted that since the re-mapping technique is applied to a task that is not started yet, and also, the technique is done in parallel with changing the frequency, the migration overhead does not affect the deadline constraints. The reason is that the latency of re-mapping is much less than the latency of changing the frequency, which we study in detail in Sect. 7.5.

7.3.3 Run-Time Management Algorithm

The pseudo-code of our proposed algorithm is outlined in Algorithm 7.1. At first, the algorithm gets the set of precedence constraint tasks, the number of tasks looking ahead (k), the scheduling table for each mode, and available *V-f* levels for cores as inputs. Then it gives the start time and the *V-f* level assignment for each task at run-time. At the initialization step, the system starts its operation in the LO mode, and also, the voltage and frequency of each core are set to the maximum value (lines 1–3). The proposed online peak power reduction algorithm is presented in lines 4–45. At first, the algorithm checks that whenever the execution time of a task exceeds its *WCET*, the system switches to the HI mode (lines 5–9). If any task execution exceeds its $WCET_i^{LO}$ and the output of this task is not ready, the system switches to the HI mode and remains in this mode till the end of the period. In this situation, in the beginning, the *V-f* level of each core is set to the maximum value to meet the deadline of HC tasks (lines 7–8). The rest of the algorithm is executed in both modes.

If there is a dynamic slack during run-time, the algorithm selects the appropriate task to assign slack, which has more impact on instantaneous power consumption (lines 10–39). This dynamic slack is generated if a task finishes its execution before its defined WCET ($WCET_i^{LO}$ or $WCET_i^{HI}$ due to the system mode). In addition, since we use static scheduling of tasks for both modes and do not change the order of task execution in each core, there may be some idle time in a core that can be used. Therefore, if there is dynamic slack, we first compute the amount of available slack (lines 11–15). Hence, we have to consider the timing overheads of the scheduler and speed changing. Therefore, we deduct these latencies from slack to guarantee the deadline (line 17). Now, we select the appropriate task among k tasks that can be released early due to the slack time after reclaimable slack (lines 18–23) based on the cost function (Eq. (7.3)). Besides, Fig. 7.2 details this process in the flowchart.

Algorithm 7.1 Online peak power reduction algorithm

Input: Task Graph (G_T), Cores, Scheduling Tables of each Mode (Sch_L and Sch_H), Number of Tasks Looking Ahead (k).

mode ← 0 , MO ← LO; // the system starts from the LO mode and Sch_L is used to schedule the tasks
for each core *j* **do**
    initialize the *V-f* level to maximum;
end for
procedure MCS OnlinePPReduction ()
    if each Task executes more than $WCET^{MO}$ **then**
        mode ← 1, MO ← HI; // System switches to the HI mode and task scheduling is done by Sch_H
        for each core *j* **do**
            initialize the *V-f* level to maximum;
        end for
    end if
    if each Task finishes its execution earlier than its deadline **or** there is an idle time in a core **then**
        if $Task_i$ has already finished its execution **then**
            S_{dyn} ← Extract_Dynamic_Slack(); //$WCET_i^{MO}$ - AET_i
        else if there is an idle time in a core
            S_{dyn} ← Extract_Dynamic_Slack(); //idle time
        end if
        $T_S, T_P \leftarrow 0$
        S_{dyn} -= TO_{sch}+TO_{Vf};
        for n = 1 **to** k **do**
            $T_P \leftarrow \tau_{n_{th}}$ after generated slack;
            if $CF_{T_S} < CF_{T_P}$ **and** T_P can start earlier **then**
                $T_S \leftarrow T_P, n_s \leftarrow$ n;
            end if
        end for
        if $n_s > 0$ **then**
            $Freq_{T_S}^{MO} \leftarrow \max\left(f_{min}, \frac{WCET_{T_S}^{MO}}{WCET_{T_S}^{MO}+S_{dyn}}\right)$
            for n = 1 **to** n_s **do**
                Update the $St_{Task_{LA-n}}^{MO}$ & $d_{Task_{LA-n}}^{MO}$
            end for
            /*Re-Mapping Checking*/
            $Cores \leftarrow Core_{T_S}, Flag_{remap} \leftarrow 0$
            for each core j in the cluster **do**
                if $CF_j < \Gamma \times CF_{Cores}$ **and** free slack exists **then**
                    $Cores \leftarrow C_j, Flag_{remap} \leftarrow 1$;
                end if
            end for
            if $Flag_{remap}$ == 1 **then**
                Re-Map (T_S, $Cores$);
            end if
        end if
    end if
    for each task *i* **do**
        if $St_i^{MO} == Time_{sys}$ or a task finishes its execution **then**
            DVFS (Ready and Running Tasks, Cores); //Update cluster *V-f* level (Algorithm 7.2)
        end if
    end for
end procedure

After determining the appropriate task, according to the system mode situation, the frequency of the core to execute the appropriate task is obtained according to the amount of slack (line 25). Hence, the selected frequency must be rounded to the nearest *V-f* level of the cluster that is greater. If there is at least one task between the generated slack and the selected task, we change their start time. Therefore, their deadline would be changed (lines 26–28). Now, the re-mapping technique is applied if the core in which the task has been allocated has a higher temperature than other cores (lines 29–37). As a result, it is possible to re-map the selected task to a core according to the cost function (Eq. (7.7)). As mentioned in Sect. 7.3.2, we just re-map a task between the cores of each cluster. Further, the algorithm checks regularly that if a task is ready to start based on the static schedules, the *V-f* level of the core that the task has been mapped on it is changed according to the defined frequency scaling factor (lines 40–44). The detail is discussed in the following subsection.

7.3.4 DVFS Governor: Updating V-f Levels in Clustered Multi-core Platform

After finishing a task execution, there might be a free slack or a task in the core queue that is ready to start its execution. Here, Algorithm 7.2 is executed to change the *V-f* level if needed. As mentioned, all cores within a cluster operate at the same *V-f* level in clustered multi-core platforms. Since the *V-f* levels of both clusters are different, it is checked on which cluster the recently completed task was running (lines 2–4). Then, we check the assigned *V-f* level of running or ready tasks on all cores of the cluster. Since all cores run with the same speed, we find the best frequency to set to the frequency cluster (lines 5–10). The reason for selecting the

Algorithm 7.2 DVFS governor

1: **function** DVFS(Ready and Running Tasks, Cores)
2:     **if** $Core_{Task_i} \leq 3$ **then**
3:         $C_{ID} = 0$; //Cluster with LITTLE cores
4:     **else**
5:         $C_{ID} = 4$; //Cluster with big cores
6:     **end if**
7:     $SetFreq = 0$;
8:     **for** $c = C_{ID}$ **to** $C_{ID} + 3$ **do**
9:         **if** $SetFreq < Freq^{MO}_{Run/ReadyTaskonCore_{C_{ID}}}$ **then**
10:             $SetFreq = Freq^{MO}_{Run/ReadyTaskonCore_{C_{ID}}}$;
11:         **end if**
12:     **end for**
13:     **if** $SetFreq \mathrel{!}= Freq_{Cluster}$ **then**
14:         `cpufreq-set -c` $Core_{Task_i}$ `-f` $SetFreq$
15:     **end if**
16: **end function**

greatest minimum frequency is to ensure that all tasks finish their execution before their deadline. In the end, if the chosen frequency (*SetFreq*) is different from cluster frequency, we change the speed of the cluster by assigning the new speed to one core of the cluster by using ⟨cpufreq-set⟩ program (lines 11–13). It should be mentioned that by changing the frequency of a cluster, its voltage will be changed automatically based on the table setting of the kernel.

7.4 Run-Time Scheduler Algorithm Optimization: Analysis and Implementation

The timing overheads of run-time scheduling and changing the frequency can have a profound impact on power-aware run-time scheduling of tasks and must be considered in the respective analysis. Neglecting them may lead to missing deadlines for MC tasks, which may cause catastrophic consequences. Two sources of generating overheads that deal with the online scheduler are the Look-Ahead Unit to select the appropriate tasks and the Re-mapping Unit to find the appropriate core. In the following, we use the online scheduler phrase for both units to make it easy to follow. The other source of causing overhead is the DVFS Governor Unit for changing frequency during run-time. Now, we first analyze the scheduler function from the timing overhead aspect. Then, we focus on optimizing the code and reducing the overheads.

In order to evaluate the scheduler and analyze the overhead on a real platform, we first convert MATLAB code to C code. Then, we detect the main parts of the code, which have more latency, and attempt to optimize it. To analyze the main functions, we first get a strict upper bound of the latency in different parts of the online scheduler on a real platform. We use the KCachegrind tool [6] to measure the worst-case time. KCachegrind is a visualization tool that uses a technique called profiling, which gives you the time distribution among the scheduler code at run-time. Now, we focus on the functions code and its timing analysis and endeavor to reduce the estimation cycles and the delay caused by cache misses in shared cache levels. Both Look-Ahead and Re-mapping Units in the online scheduler are called frequently during run-time, and the apparent improvement and optimization should be performed. The run-time phase of Fig. 7.2 shows the flowchart of these two units in detail. As shown in this figure, some functions play a critical role in the timing overhead of the power-aware run-time scheduler, which is indicated by the white color. As discussed in the previous section, the Look-Ahead Unit chooses k tasks after generating dynamic slack and finds a task that has the most effect on peak power and maximum temperature. Checking the k tasks is done in a for-loop, in which each task is investigated that can release early to use the dynamic slack. Therefore, all predecessors of it must be checked whether they can finish their execution soon or not. Investigating the execution status of all predecessors needs more cycles to be done and then causes latency and more cache misses. Therefore, introducing an entity that shows the estimated finish time of a task would be useful, and instead of checking the status of all predecessors, just that entity can be checked. Besides, due to the having different *V-f* levels, we must ensure that the dynamic slack is large enough to include the timing overhead of changing the *V-f* level. This check prevents the over-calling of the Re-mapping Unit. In addition, there are two functions for calculating the costs, in which there are some math calculations with high timing overhead. Hence, optimizing these calculations by predefining them to avoid dynamic memory allocation during computation would help reduce timing overhead.

From the perspective of cache hit/miss, one of the ideas is to optimize the code to reduce the estimation cycles in the online scheduler by focusing on calculations and memory access latency. There are some tips to optimize C/C++ code to run it faster: reducing function calls and the number of function parameters, how to define variables and objects, how to use operators, using prefixes instead of postfixes in objects, avoiding unnecessary data initialization, and so many other techniques that we must use for code optimization. Apart from using these techniques, due to data access latency, we have effective timing overhead in the online scheduler. We optimize code by changing the representation of the data structure manipulated by the algorithms. We have defined two types of task classes: (1) defining a task class that uses vectors in the class for each task entity and (2) defining a task class with vectors of task class in the number of tasks. Each has its advantages and disadvantages under certain circumstances. However, due to the checking of limited tasks (k) in the run-time scheduler, the use of the second task class has less timing overhead and cache misses. As a result, Table 7.1 shows the percent of cache (L1 and LL (last level)) read and write misses for Look-Ahead and Re-mapping Units after optimization on three different platforms, Cortex A7, Cortex A15, and Intel Core i5. As mentioned in previous sections, most of the embedded systems use ARM processors, not Intel. Therefore, we target the ARM processors, such as the ODROID board. However, this table shows that we have less than 3.584% and 0.081% cache L1 and LL data misses in ARM processors, respectively, which are admissible compared to all cache misses and also in comparison with Intel processor that has fewer cache misses.

Table 7.1 Run-time scheduler cache misses' report

	Look-Ahead Unit			Re-mapping Unit		
	Cortex A7	Cortex A15	Intel Core i5	Cortex A7	Cortex A15	Intel Core i5
L1 Data Read Miss	3.318%	3.318%	2.946%	0.163%	0.163%	0.303%
L1 Data Write Miss	3.584%	3.584%	2.168%	0.352%	0.352%	0.764%
LL Data Read Miss	0.076%	0.081%	0.0%	0.028%	0.034%	0.0%
LL Data Write Miss	0.063%	0.0%	0.0%	0.036%	0.0%	0.0%

7.5 Evaluation

7.5.1 Experimental Setup

7.5.1.1 Hardware Platform

To evaluate our system, we conducted experiments on the ODROID XU3/XU4 board powered by ARM, which has a big.LITTLE architecture, four big (Cortex A15), and four LITTLE (Cortex A7) cores. The ODROID XU3 board supports DVFS and can operate at 13 different *V-f* levels between [$0.9V$, 200 MHz] and [$1.3V$, 1.4 GHz] on LITTLE cores, while the last four frequency levels have the same voltage levels and 19 different *V-f* levels between [$0.9V$, 200 MHz] and [$1.3625V$, 2 GHz] on big cores. Therefore, the effect of changing *V-f* levels is done by scaling the frequency within the range of available levels.

7.5.1.2 Task Set Generation

In the experiments, we use random applications (task graphs) generated by the tool in [4]. An example of a real-life application is already given in the motivational example. In these applications, there are four basic parameters, c (number of cores), U (system utilization), d (outgoing edge percentage), and n (number of tasks), which are presented in Table 7.2. d represents the probability of having outward edges from one task to the others. In addition, U/c is the normalized system utilization that refers to both LC and HC tasks with their predefined $WCET^{HI}$. As the results are presented in both simulation and real platform (with eight cores), we show the results with 16 cores in simulation in addition to 2, 4, and 8 cores. We provide different configurations by changing the value of these parameters for different scenarios used in the experiments.

7.5.1.3 Tasks' Power Consumption

In order to have a realistic possible range of power values, we run several embedded benchmarks from MiBench suite [7], e.g., automotive, network, and Telecomm. benchmarks on two configurations, ARM Cortex A7 and A15 on the ODROID XU3

Table 7.2 Experiment configurations

Param.	Varying c	Varying U/c	Varying n	Varying d
c (#core)	2, 4, 8, 16	8	8	8
U/c (utilization)	[0.5, 0.75]	[0, 1]	[0.5, 0.75]	[0.5, 0.75]
d (edge percentage)	10%	10%	10%	1%, 10%, 20%
n (#task)	50	50	30, 40, 50, 80	50

with maximum frequency and read data from power sensors on the board. Hence, since the DVFS is applied to the whole processor, the power consumption at other lower frequencies can be obtained using Eq. (2.4) in Sect. 2.1.4 by considering frequency scaling [8]. In addition, we examined different scenarios of activating one core to all cores by running different benchmarks. We run each benchmark 1000 times and report the maximum value of power consumption. We select the maximum power of tasks in the range of these minimum and maximum values in our experiments, which is $[484, 940]mW$ in Cortex A7 and $[3.891, 7.622]W$ in Cortex A15. The power that the tasks may consume is generated randomly following the normal distribution within this range. Besides, we consider the power consumption of the system as the sum of the power consumption of all cores [9].

7.5.1.4 Thermal Analysis

As presented in the proposed method section, we assume that our approach does not have to probe the core temperature to make a decision. Therefore, during the scheduling of the tasks, the power values of cores depending on the running tasks are recorded. In addition, for validating on the real platform, since there are just temperature sensors for big cores on the ODROID XU3, the HOTSPOT tool [10] is used to obtain the core temperature throughout the execution for the specific floorplan and configuration platform which we use. For the configuration file, we use the parameters reported in [11], which is for ARM big.LITTLE processors. The ARM core (A7) has an area of 0.45 mm^2 in our experiments reported by the ARM company.

7.5.1.5 Comparison

In this chapter, we analyze the proposed method and compare it against [4, 12]. The work [4] proposes an offline scheduling algorithm for an MC system where most of the LC tasks are not dropped in the HI mode to improve the QoS of the system. However, they ignore the peak power and temperature aspect of the system. Additionally, researchers in [12] suggest an online energy minimization algorithm for hard real-time systems where they use the dynamic slack just for the immediately available task to decrease the *V-f* level. We consider the latency while comparing with the method of [12] to have a fair comparison.

7.5.2 Analyzing the Relevance Between a Core Temperature and Energy Consumption

At first, we represent the relevance between core temperature and its accumulated energy consumption. In Sect. 7.3.2, our algorithm was based on the assumption that

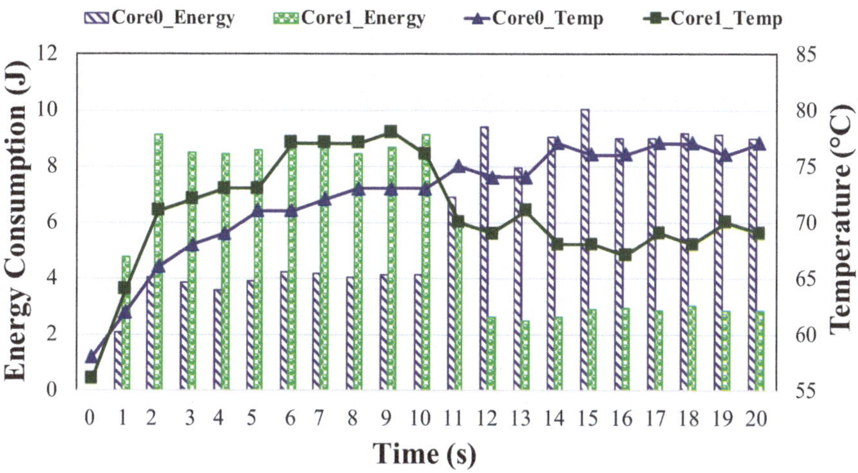

Fig. 7.5 The relevance between a core temperature and accumulated energy consumption

a core tends to have a lower temperature when its accumulated energy is less than the other cores. Figure 7.5 studies the validity of the assumption. Since we do not have a power sensor for each big core on the ODROID XU3, we run the same task on all the big cores to have the same power consumption. This task is executed several times periodically in cores with different execution times. Therefore, we have different energy consumption in each period of cores. After finishing the execution of the task on each core, the core goes to sleep until the end of the task period. Figure 7.5 shows energy consumption and the temperature of two big cores during the time for a window of energy monitoring equal to two seconds. In this figure, first, the task runs with a larger execution time on Core1 in comparison to Core0. Thus, the temperature of Core1 rises more rapidly than Core0. After 10s, the accumulated energy of Core1 is reduced, and Core0 is increased. As shown, Core0 that has more energy consumption tends to have a higher temperature.

7.5.3 The Effect of Varying Parameters of Cost Functions

Now, we evaluate the results for different values of α and β in Eq. (7.3). The experiments are carried out for a system with $c = 8$, $U/c \in [0.5, 0.75]$, $d = 1\%$, and $n = 30$. The average results (Fig. 7.6) are obtained for a set of 100 task graphs with different $\langle \alpha, \beta \rangle = \langle 0, 1 \rangle$, $\langle 0.25, 0.75 \rangle$, $\langle 0.5, 0.5 \rangle$, $\langle 0.75, 0.25 \rangle$, and $\langle 1, 0 \rangle$. The results are normalized to [4]. In this section, to show the effect of varying these two parameters, tasks are executed with their actual execution time (AET), and task re-mapping is not exploited. It can be seen that, in every case, utilizing our approach would lead to a system with lower peak power, energy, and peak temperature.

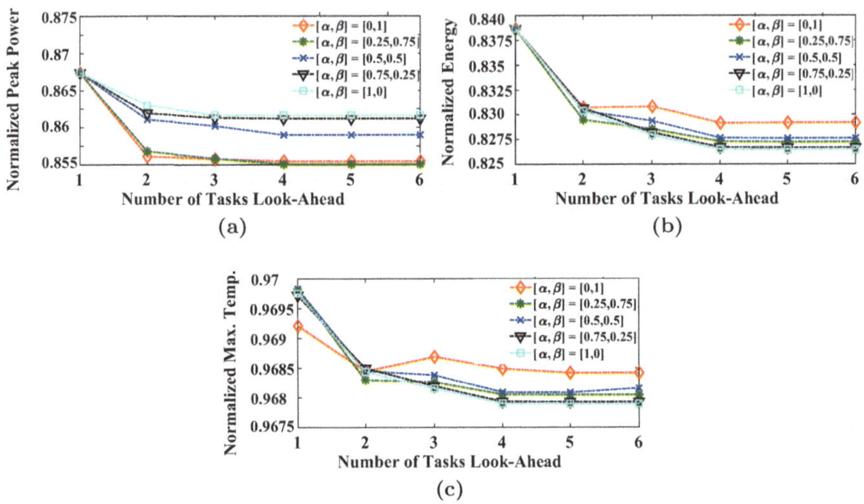

Fig. 7.6 Impact of varying α and β on peak power, energy, and maximum temperature. (**a**) Normalized peak power. (**b**) Normalized energy. (**c**) Normalized peak temperature

Besides, the expected effect of varying $\langle\alpha, \beta\rangle$ is confirmed in the experiments. For example, the average normalized peak power is progressively reduced when β increases from 0 to 1, as presented in Fig. 7.6a. Similarly, in Fig. 7.6b, the higher the α, the lower the energy consumption and peak temperature. Finally, as the algorithm looks further ahead in the future to find the best tasks to assign the dynamic slack, the results are generally getting better, up to 1.25% and 1.25% more reduction in peak power and energy. It is worth noting that, in this experiment, we intentionally disable the task re-mapping technique to ensure that the effect of $\langle\alpha, \beta\rangle$ is not skewed by another optimization.

For the other experiments in this work, we consider $\langle\alpha, \beta\rangle = \langle 0.5, 0.5\rangle$ that balances both peak power and temperature average reduction in comparison with other values of $\langle\alpha, \beta\rangle$.

7.5.4 The Optimum Number of Tasks to Look-Ahead and the Effect of Task Re-mapping

In this subsection, we analyze the optimum number of tasks to look-ahead (k) by evaluating the respective average quality of results without considering the overheads. The number of look-ahead tasks is varied from 1 to 10. The results presented in Fig. 7.7 are obtained from some scenarios of changing parameters in Table 7.2 with running on a homogeneous multi-core system. In scenarios, we have considered the following changes for evaluation shown in Fig. 7.7: $n = 80$ when

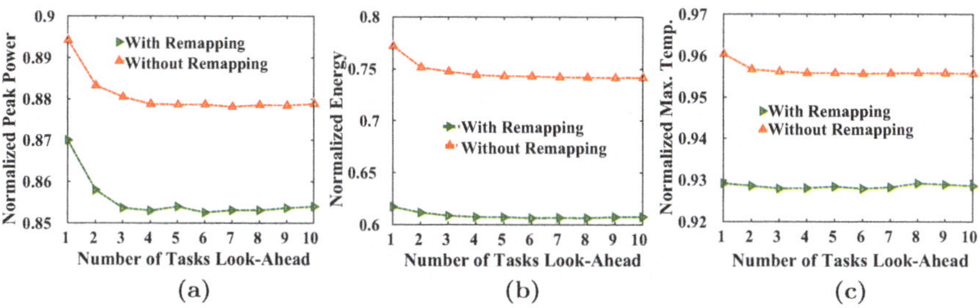

Fig. 7.7 Normalized improvement in peak power, energy, and maximum temperature for all scenarios. (**a**) Peak power. (**b**) Energy. (**c**) Max. temperature

c is varied, $c = 4$ when n is varied, and $n = 40$ and $c = 4$ when d is varied. As a result, looking four tasks ahead provides a significant reduction in peak power and also in maximum temperature and energy consumption with and without task re-mapping. Hence, looking four tasks ahead is the average result of varying all parameters. Besides, when task re-mapping is used, the temperature, on average, is reduced by 2.7% (2 °C) compared to the case where task re-mapping is disabled. In general, by looking ahead four tasks and enabling task re-mapping, the proposed method reduces the peak power, energy consumption, and maximum temperature on average by 14.6%, 39%, and 7.1% (6.1 °C), respectively, compared to [4] and 4.2%, 16%, and 3.1% (2.7 °C), respectively, compared to [12]. Hence, looking four tasks ahead is the average result of varying some parameters. The detail of finding the optimum k by varying the properties of tasks is discussed as follows.

We show the relation between the number of look-ahead tasks (k) and the task property, edge percentage ($d\%$) to model the system capability such as peak power minimization, energy consumption, and maximum temperature. For this analysis, the data from the experiments with four cores (c) and the system utilization per core (U/c) in the range of [0.5,0.75) is used. The average data of 100 task set runs has been used.

We use the MATLAB Curve Fitting Tool to derive the polynomial functions of various system parameters. Figure 7.8 shows the curve of the system peak power consumption by varying k and d normalized to the result for $k = 1$, and the corresponding equation is shown in Eq. (7.8). This equation is the polynomial function with the maximum degree of four with the minimum Root Mean Square Error (RMSE), equal to 0.0024:

$$\begin{aligned} Pow_{peak}^{Norm.}(k,d) = 1.033 - 0.5082d - 0.03374k + 1.332d^2 + \\ 0.1799dk + 0.006184k^2 - 0.7267d^2k - 0.01158dk^2 - \\ 0.0005375k^3 + 0.05294d^2k^2 - 0.0001314dk^3 + 1.912 \times 10^{-5}k^4 \end{aligned} \tag{7.8}$$

The equation above can also be used to mathematically derive the optimal k for a particular task property to optimize the various metrics. For example, if d is kept

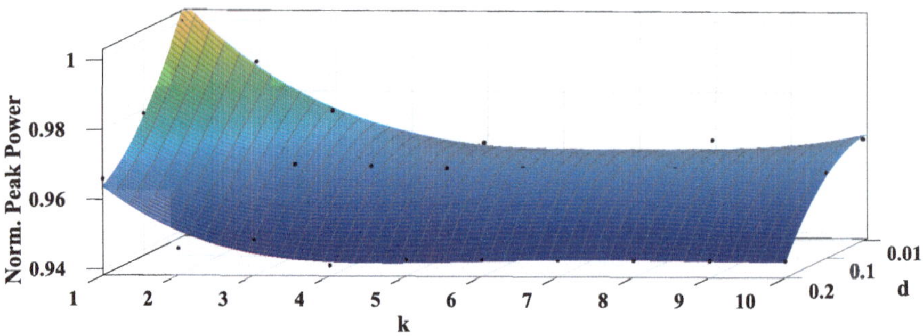

Fig. 7.8 Impact of number of look-ahead tasks and edge percentage on normalized peak power consumption

Fig. 7.9 Impact of number of look-ahead tasks on normalized peak power, energy, and maximum temperature while $d = 20\%$

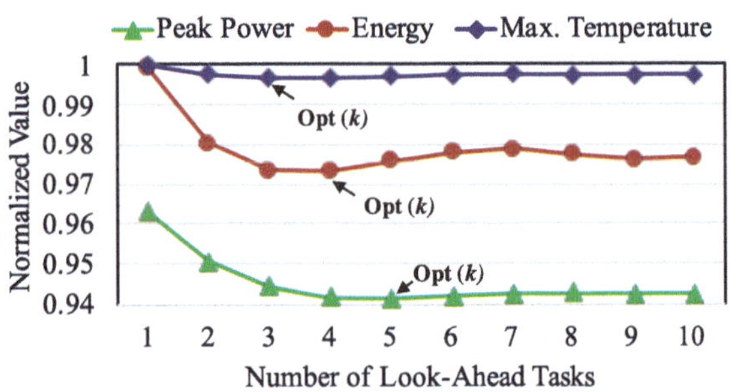

as 20% in Eq. (7.8), the minimum value of the curve is obtained when $k = 5$, which is shown in Fig. 7.9. In addition, by deriving the corresponding equations for the normalized maximum temperature (Eq. (7.9)) and energy consumption (Eq. (7.10)), the optimum value of k for the system's maximum temperature is $k = 3$ and for the energy consumption is $k = 4$, as shown in Fig. 7.9:

$$Energy_{peak}^{Norm.}(k,d) = 0.9347 + 1.176d - 0.05964k - 3.304d^2$$
$$-0.01962dk + 0.01355k^2 + 0.3278d^2k + 0.005128dk^2 -$$
$$0.001431k^3 - 0.02466d^2k^2 - 0.000231dk^3 + 5.616 \times 10^{-5}k^4 \quad (7.9)$$

$$T_{peak}^{Norm.}(k,d) = 0.9925 + 0.07826d - 0.006975k - 0.08123d^2$$
$$-0.008346dk + 0.001644k^2 + 0.05333d^2k + 0.001238dk^2$$
$$-0.0001762k^3 - 0.003814d^2k^2 - 10^{-5}(4.653dk^3 + 0.6901k^4) \quad (7.10)$$

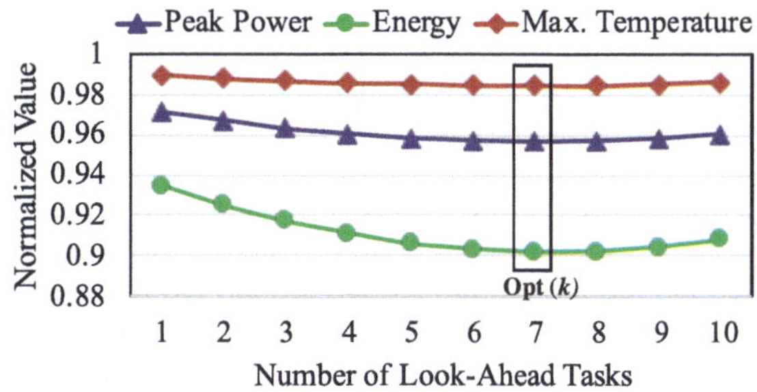

Fig. 7.10 Impact of number of look-ahead tasks on normalized peak power, energy, and maximum temperature with consideration of the overheads

Now, we analyze the optimum number of look-ahead tasks by evaluating the respective average results, considering the overheads and no task re-mapping. The results presented in Fig. 7.10 are obtained from the scenarios of varying c, n, and d parameters in Table 7.2 by all mentioned values. In general, this figure shows that the objectives (peak power, energy, and maximum temperature) are improved by increasing the number of look-ahead tasks. However, the efficiency of our methods is outstanding when there is much slack at run-time; therefore, by increasing the number of look-ahead tasks, the timing overheads of the Task Selection Unit (for determining the appropriate task in objective improvement) are more. Therefore, it causes less dynamic slack for assigning it to the appropriate task, leading to less improvement in objectives. According to this figure, looking seven tasks ahead significantly reduces peak power, maximum temperature, and energy consumption. Thus, since overhead is considered in the evaluation, seven tasks are considered the optimal number to look ahead in the rest of this chapter.

7.5.5 The Analysis of Scheduler Timings' Overhead on Different Real Platforms

To investigate the timing overhead of the proposed run-time scheduler, we analyze it on three real platforms, Intel Core i5, ARM big core (A15), and ARM LITTLE core (A7) on ODROID XU3/4 and ARM core (A53) in Xilinx Zynq UltraScale+ MPSoC board. Figure 7.11 shows the overheads in each platform for different numbers of tasks looking ahead. Each boxplot shows the average latency for normal values of parameters in Table 7.2 ($d = 10\%$, $c = 8$, $U/c = [0.5, 0.75]$, $n = 80$) with 100 task graphs. The following observation can be seen from the figure. First of all, the run-time timing overhead in the Intel platform is extremely small as compared to ARM processors. In the second observation, as the number of task look-ahead is increased, the latency is increased in all ARM processors. However, this latency increase is more evident in the A7 processor, while it is almost constant after looking seven tasks ahead in the A53 processor and four tasks ahead in the A15 processor. Furthermore, since big (A15) cores have high performance as compared to LITTLE (A7) cores, this timing overhead would be less. However, since a platform has lower

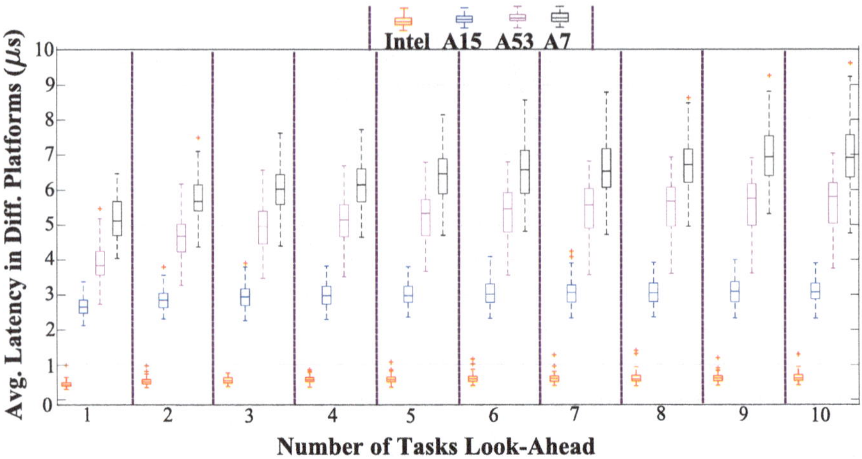

Fig. 7.11 Analyzing the timing overhead of run-time scheduler on four platforms

performance, the range of latency between the minimum and maximum value is more significant. This fact is due to its lower performance and access to the memory and cache miss/hit.

To have a real implementation of our proposed method, we obtain the observed worst-case timing overhead of the run-time scheduler, in which the appropriate task is selected for slack assignment and also a proper core for the re-mapping. Since many embedded systems use ARM processors, we evaluate our method on the ARM processor. We analyze the overhead on a LITTLE core of the ODROID XU4 platform. We examine both the Look-Ahead Unit and Re-mapping Unit in the run-time scheduler separately and obtain the maximum observed timing overhead. To determine this overhead, we ran several applications (200 task graphs) with their various inputs on ARM LITTLE Core (A7). Based on these overheads, we set the observed worst-case of the Look-Ahead Unit to the maximum value. In addition, the scheduler in Re-mapping Unit checks other cores, whether it is possible to re-map a task for thermal management. In the worst case, all cores are checked. Therefore, to have a close to accurate observed worst-case timing overhead of the Re-mapping Unit, we obtain the worst timing overhead for each core and multiply it by the number of cores. Based on our observation and this measurement, the maximum timing overhead for Look-Ahead Unit is 56.417 μs, and for Re-mapping Unit per core is 64.54 μs. In order to ensure that the timing guarantees provided by the static schedule are not violated, we deduct these overheads from slack before assigning it to an appropriate task.

7.5.6 The Latency of Changing Frequency in Real Platform

The main unit, the DVFS governor, adjusts the frequency, which has a significant timing overhead. The ODROID XU3/4 board has a frequency range of ⟨0.2, 1.4⟩ GHz for LITTLE cores and ⟨0.2, 2⟩ GHz for big cores, with the step of 0.1 GHz. We change the frequency by using the ⟨cpufreq-set⟩ program in two scenarios of scaling-down and scaling-up. Hence, the voltage is adjusted automatically according to the selected frequency. The maximum latency for all scenarios is at most 12.025 ms. Besides we observe in our experiments that the latency of the scaling-down transition is 342 μs less than the scaling-up transition, on average. Regardless of the frequency scaling-down or up, we consider the latency of changing *V-f* level to be equal to 12.025 ms. Due to this timing overhead, we deduct this latency from available dynamic slack before assigning it to a task to guarantee the correct execution of tasks before their deadlines. Since the re-mapping latency is $3.75ms$ [13] in the worst-case scenario and re-mapping is done in parallel with changing the frequency, it has no impact on the overall deadline.

7.5.7 The Effect of Latency on System Schedulability

As discussed, considering the latencies of the run-time scheduler and changing frequency are critical in analyzing the system. If these timing overheads are not studied, it may cause deadline miss of tasks and then catastrophic consequences. Our proposed method's effectiveness depends on the available slacks at run-time and the possibility of assigning them to the tasks. Therefore, if the latencies are not properly accounted for, some tasks may not be executed successfully before their deadline. Figure 7.12 shows the percentage of successfully executed task sets

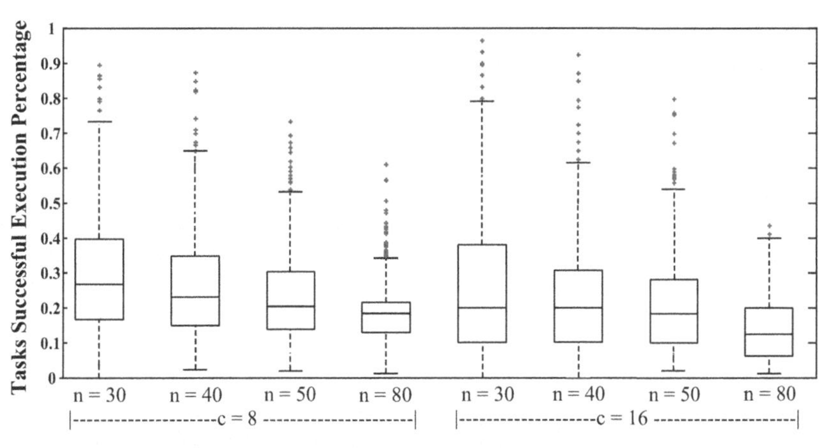

Fig. 7.12 Percent of successful executed tasks before their deadline in task graph according to our proposed method without considering timing overheads

before their deadlines during the run-time phase in different scenarios if we do not consider the timing overheads. The results are obtained for the normal scenario of some parameters ($d = 10\%$, $U/c = [0.5, 0.75]$, $c = 8$, 16, and $n = 30, 40, 50$, and 80) and 1000 task graphs for each scenario. In this figure, we observe that as the number of tasks in the system with the same number of cores is increased, fewer task sets can be scheduled and meet their deadline. When there are more tasks in the system with the same *U/c*, the dynamic slacks that are incurred when the tasks finish earlier than their WCETs are smaller. The reason is that as the expected execution times of the tasks are decreased, the absolute differences between their AET and WCETs are inherently small. Therefore, the possibility of missing a deadline is increased, and fewer tasks would be executed successfully before their deadline. In addition, if the number of cores in the system is increased, more task sets miss their deadlines. Since the re-mapping technique is used to manage temperature, all cores are checked in the worst case. Therefore, by increasing the number of cores, the timing overhead of selecting a proper core for task re-mapping is increased. Since this latency has not been considered while a dynamic slack is assigned, using the re-mapping technique at run-time may cause more deadline violations by increasing the number of cores. In general, as shown in this figure, a high percentage of task sets miss their deadline, which is not acceptable in MC systems. Therefore, it is critical to consider the timing overheads of the run-time scheduler and changing frequency.

7.5.8 The Analysis of the Proposed Method on Improving Objectives in Simulation

In order to illustrate how effective our proposed method is with different parameters, we analyze the results under four separate scenarios of Table 7.2, shown in Fig. 7.13, in which the results are normalized to [4]. These results are obtained for multi-core systems, in which there are homogeneous cores based on ARM A7. In general, as the applications get more complicated (e.g., having a large number of tasks or system utilization), it is harder to achieve significant savings in peak power, energy, and maximum temperature. Thanks to our task re-mapping technique, where the tasks are redistributed more evenly to the cores at run-time based on their accumulated energy, the maximum temperature is properly managed.

For the case of varying the number of cores, since our method only tries to optimize the peak power for each core individually to reduce the time overhead, it is more difficult to maintain a similar system peak power reduction when c is low. Nevertheless, as illustrated in Fig. 7.13a, the difference in peak power is significant by increasing the number of cores. In addition, as the temperature of each core is affected by the temperatures of neighboring cores, the reduction in maximum temperature is less by increasing the number of cores. On average, the

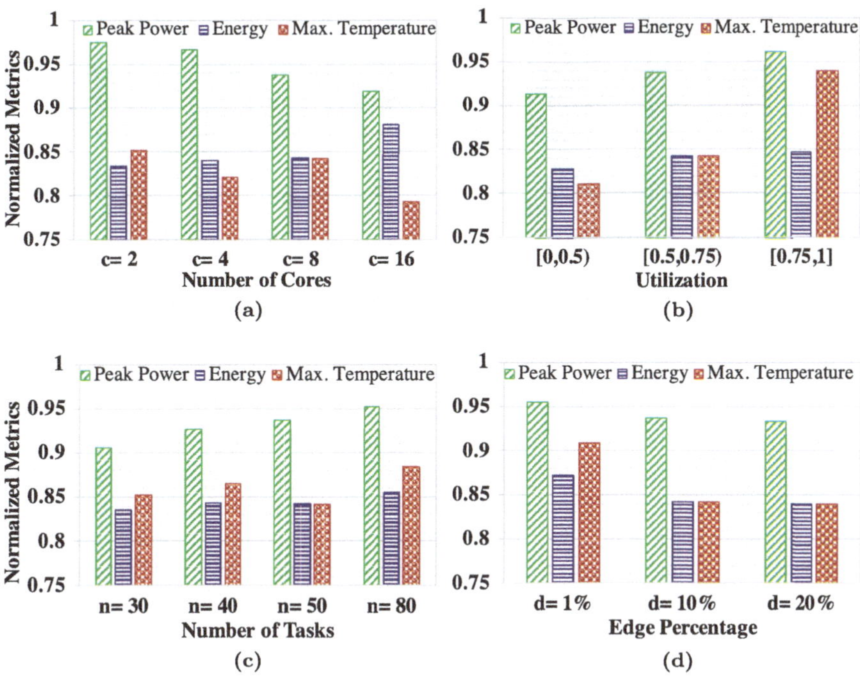

Fig. 7.13 The improvements in peak power, energy, and maximum temperature in different scenarios normalized to [4]. (**a**) Varying number of cores. (**b**) Varying utilization bound. (**c**) Varying number of tasks. (**d**) Varying edge percentage

peak power, maximum temperature, and energy consumption in the system are reduced by 5.015%, 14.12 °C, and 15.073%, respectively.

The effectiveness of our method depends on the available slacks at run-time and the possibility of assigning them to the tasks. Therefore, if there is less slack due to the nature of the application in terms of the number of tasks and system utilization, the reduction in peak power, energy consumption, and maximum temperature is less. For instance, in Fig. 7.13b, when the system utilization is getting higher, the idle time of the core between two consecutive tasks is getting smaller. The tasks also tend to execute longer. Thus, the amount of slacks that can be exploited at run-time is limited. But, overall, the peak power is reduced by at least 3.905% and up to 8.59% in this scenario. Similarly, when there are more tasks in the system with the same U/c, the dynamic slacks incurred when the tasks finish earlier than their WCETs are smaller. The reason is that as the expected execution times of the tasks are decreased, the absolute differences between their AET and WCETs are inherently small. However, as seen in Fig. 7.13c, our method manages to reduce the peak power, energy, and maximum temperature on average by 6.96%, 15.61%, and 10.54 °C, respectively.

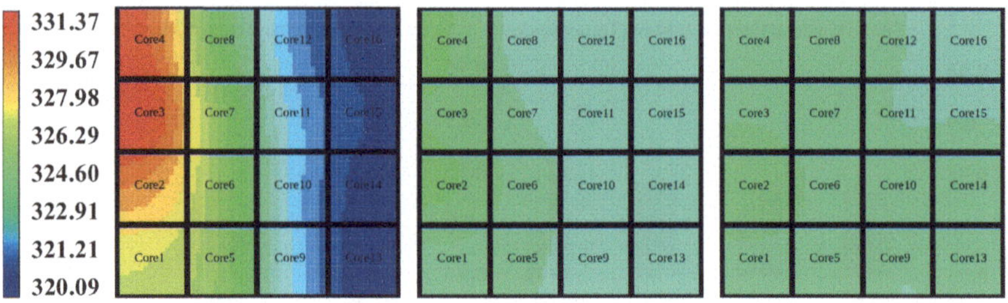

Fig. 7.14 Temperature profile for different edge percentages. (**a**) Medina et al. [4] with $d = 20\%$. (**b**) $d = 20\%$ and $k = 1$. (**c**) $d = 20\%$ and $k = 4$

Besides, the possibility of releasing the tasks earlier than their presumed start times also affects the outcomes. When the dependency between the tasks is high, a significant amount of them cannot be released earlier. This behavior can have either a positive or negative impact on the system. For the former, the cores might have more idle time because the tasks have to wait longer for their predecessors to finish. For the latter, our method has less opportunity to apply DVFS to tasks. However, at run-time, these idle periods might overlap with the other tasks with the already reduced *V-f* level. The peak power of the whole system is consequently reduced. It can be seen in Fig. 7.13d that when $d = 20\%$, the best peak power and maximum temperature reduction are achieved compared to the cases where $d = 1\%$ and $d = 10\%$.

Figure 7.14 shows an example steady-state heat map of the systems with different edge percentage parameters. The result is obtained for a system with $c = 16$, $U/c = 0.5$, and $n = 80$. We show the results of looking one and four tasks (which is the optimum value when we do not consider timing overhead) ahead as compared to [4]. It can be observed that our approach not only reduces the maximum temperature but also helps in balancing the difference in temperature between the cores, especially when $k = 4$.

7.5.9 *The Analysis of the Proposed Method on Improving Objectives in a Platform Based on ODROID Architecture*

In this section, we analyze the improvement of peak power, energy consumption, and maximum temperature in a clustered heterogeneous multi-core processor in which there are four big (A15) and four LITTLE (A7) cores. Here, we have a common *V-f* level for all cores within the same cluster (cluster with big cores or cluster with LITTLE cores), while in the previous results, the frequency of each core was changed individually. We show the results in Fig. 7.15, in which the improvements in peak power, energy consumption, and maximum temperature

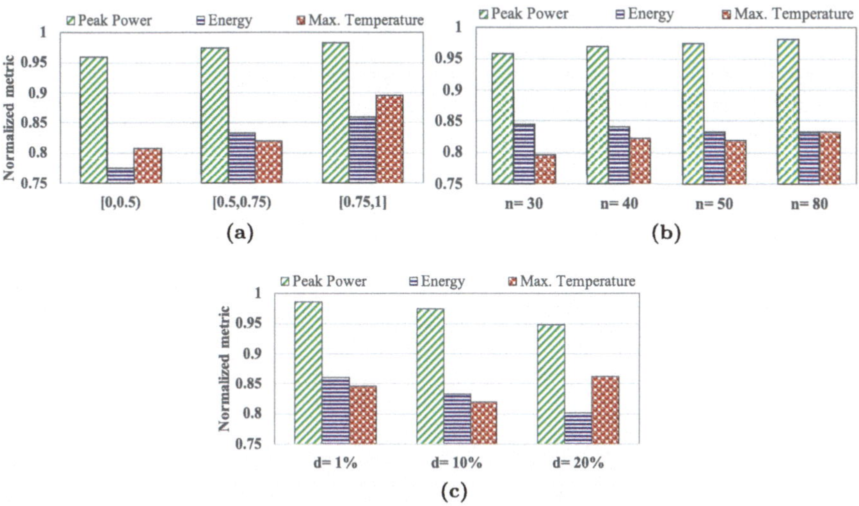

Fig. 7.15 Normalized metrics running on the clustered heterogeneous multi-core platforms. (**a**) Varying utilization bound. (**b**) Varying number of tasks. (**c**) Varying edge percentage

in the clustered heterogeneous multi-core system across all experiments are up to 5.25%, 22.44%, and 20.33% (16.8 °C), respectively, in comparison with [4]. The trends for the clustered heterogeneous multi-core architecture are similar to that obtained for the homogeneous architecture in the previous subsection. The only notable difference is in the peak power, which is on average 3.461% worse than the tasks on a homogeneous multi-core system, due to enforcing of common *V-f* level for the entire cluster. Therefore, the power improvement is somewhat lower.

7.5.10 Evaluation of Running Real MC Task Graph Model on Real Platform

Now, we validate the proposed online technique with a real-life application task graph, presented in Fig. 7.1a, running on the ODROID XU3. In particular, we evaluate the impacts of changing frequency on the system power and temperature in this section. The Unmanned Air Vehicle (UAV) application consists of seven dependent tasks executed on two cores, which has been presented in Fig. 7.1a. Since there are no available real benchmarks for the tasks of the graph, we used different benchmarks of MiBench [7] as tasks of the graph. Then, we obtain the WCETs and maximum power consumption of each task running on the ODROID XU3. Since WCET analysis is a complicated task [14], we used an existing WCET estimation tool called OTAWA [15], to capture the high WCETs ($WCET_i^{HI}$). In addition, we run each benchmark 10,000 times and select the maximum of the

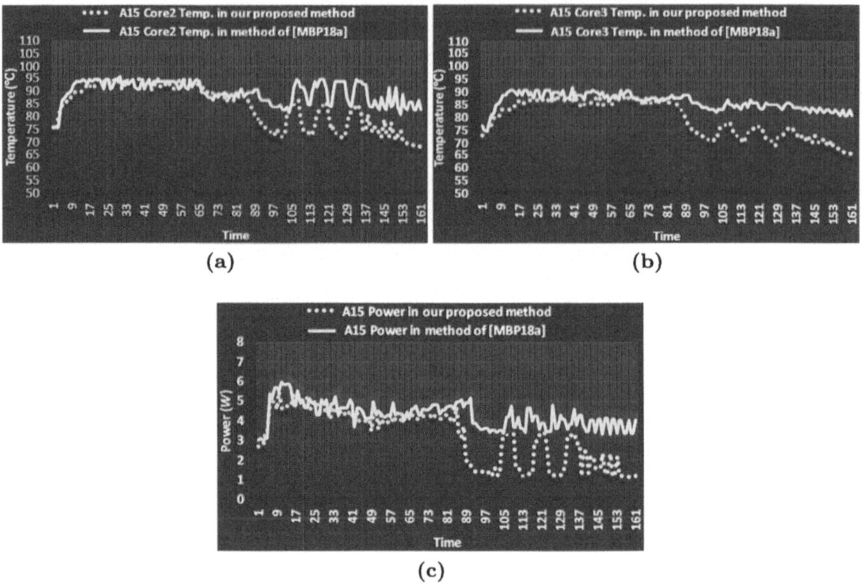

Fig. 7.16 Power and temperature sensor data of the ODROID XU3, by running the real MC task graph (unmanned air vehicle). (**a**) Temperature trace of A15-core2. (**b**) Temperature trace of A15-core3. (**c**) Power trace of the big cluster

measured execution time as the low WCET ($WCET_i^{LO}$). To analyze the system temperature, we run the application on Core2 and Core3 that in general have a higher temperature due to their proximity to the memory and other components. Hence, there is a temperature sensor for each big core and a power sensor for each cluster of ODROID XU3. Therefore, the power and temperature data of this section are exploited from the board sensors.

Figure 7.16 shows the power trace of the cluster of big cores and temperature trace of two cores during run-time in two scenarios of using our DVFS-based proposed method and presented method of [4] ([MBP18a]). At run-time, to analyze a task execution time, we select a docker container to run a task and check the time to be aware of the exact start and finish times of the task. Then, the dynamic slack is computed, the slack between the task's actual completion time and its WCET. Figure 7.16c shows the power traces of the method of [4] and our proposed method by considering two tasks looking ahead that the DVFS has been used from almost 90 ms. In addition, as shown in Fig. 7.16a and b, in general, the average core temperature has been decreased by using the DVFS-based proposed method and looking two tasks ahead. Based on the scheduling of tasks in Fig. 7.1, one of the cores is active until near the middle of the period. However, the temperature of each core is affected by the temperature of neighboring cores. In addition, after executing two tasks in each core and using the dynamic slack to reduce the speed, the cores' temperatures are decreasing. Besides, in a part of the task graph period, only Core2 is active but still has high temperatures. Therefore, after applying the proposed method and reducing the *V-f* levels, the cores' temperatures are reduced. The proposed method will be more effective and have a significant improvement if there are more tasks that are run on a system with more cores.

7.6 Conclusions

In this chapter, we studied peak power and peak temperature reduction in MC embedded systems at run-time and analyzed the proposed run-time power-aware scheduler on clustered multi-core real platforms while guaranteeing the minimum QoS value. Our presented method uses the re-mapping technique and DVFS at run-time whenever there is a dynamic slack. We also proposed the associated cost functions to select the most appropriate task to assign the dynamic slacks to decrease its *V-f* level or re-map it to another core. We showed that more peak power and maximum temperature reduction are achieved by increasing the number of tasks to look ahead. In addition, the proposed power-aware scheduler was analyzed in terms of run-time timing overhead in different multi-core platforms. We focused on reducing the run-time scheduler latency to have more usage of dynamic slack and, consequently, more peak power and maximum temperature reduction while guaranteeing the deadlines of LC and HC tasks in their specified system mode. Besides, although we consider dual-criticality level systems in this work, the proposed approach can be applied to any MC system, regardless of how many criticality levels of tasks are executed in the system. The results show up to 5.25%, 20.33% (16.8 °C), and 22.44% reduction in peak power, maximum temperature, and average energy consumption, respectively, compared to recent studies.

References

1. Mohammad Salehi, Alireza Ejlali, and Bashir M Al-Hashimi. "Two-phase low-energy N-modular redundancy for hard real-time multi-core systems". In: *IEEE Transactions on Parallel and Distributed Systems (TPDS)* 27.5 (2016), pp. 1497–1510.
2. Alireza Ejlali, Bashir M. Al-Hashimi, and Petru Eles. "Low-Energy Standby-Sparing for Hard Real-Time Systems". In: *IEEE Transactions on Computer-Aided Design of Integrated Circuits and Systems (TCAD)* 31.3 (2012), pp. 329–342. https://doi.org/10.1109/TCAD.2011.2173488.
3. Yifeng Guo, Dakai Zhu, and Hakan Aydin. "Reliability-aware power management for parallel real-time applications with precedence constraints". In: *Proc. pn International Green Computing Conference and Workshops*. 2011, pp. 1–8. https://doi.org/10.1109/IGCC.2011.6008562.
4. R. Medina, E. Borde, and L. Pautet. "Availability enhancement and analysis for mixed-criticality systems on multi-core". In: *Proc. on Design, Automation & Test in Europe Conference & Exhibition (DATE)*. 2018, pp. 1271–1276.
5. Pengcheng Huang et al. "Energy efficient dvfs scheduling for mixed-criticality systems". In: *Proc. on Embedded Software (EMSOFT)*. 2014, pp. 1–10.

6. Josef Weidendorfe. *Kcachegrind tool*. http://kcachegrind.sourceforge.net/html/Home.html. Accessed: December 2019.
7. M. R. Guthaus et al. "MiBench: A free, commercially representative embedded benchmark suite". In: *Proc. IEEE International Workshop on Workload Characterization. WWC-4*. 2001, pp. 3–14. https://doi.org/10.1109/WWC.2001.990739.
8. M. Salehi and A. Ejlali. "A Hardware Platform for Evaluating Low-Energy Multiprocessor Embedded Systems Based on COTS Devices". In: *IEEE Transactions on Industrial Electronics (TIE)* 62.2 (2015), pp. 1262–1269.
9. Waqaas Munawar et al. "Peak Power Management for scheduling real-time tasks on heterogeneous many-core systems". In: *Proc. of the International Conference on Parallel and Distributed Systems (ICPADS)*. 2014, pp. 200–209.
10. Wei Huang et al. "HotSpot: A compact thermal modeling methodology for early-stage VLSI design". In: *IEEE Transactions on Very Large Scale Integration (VLSI) Systems* 14.5 (2006), pp. 501–513.
11. Y. Gong, J. J. Yoo, and S. W. Chung. "Thermal Modeling and Validation of a Real-World Mobile AP". In: *IEEE Design & Test* 35.1 (2018), pp. 55–62.
12. D. Zhu, R. Melhem, and B. R. Childers. "Scheduling with dynamic voltage/speed adjustment using slack reclamation in multiprocessor real-time systems". In: *IEEE Transactions on Parallel and Distributed Systems (TPDS)* 14.7 (2003), pp. 686–700. https://doi.org/10.1109/TPDS.2003.1214320.
13. K. R. Basireddy et al. "AdaMD: Adaptive Mapping and DVFS for Energy-efficient Heterogeneous Multi-cores". In: *IEEE Transactions on Computer-Aided Design of Integrated Circuits and Systems (TCAD)* (2019). https://doi.org/10.1109/TCAD.2019.2935065.
14. M. Bazzaz, A. Hoseinghorban, and A. Ejlali. "Fast and predictable non-volatile data memory for real-time embedded systems". In: *IEEE Transactions on Computers (TC)* (2020), pp. 1–1.
15. Clément Ballabriga et al. "OTAWA: an open toolbox for adaptive WCET analysis". In: *IFIP International Workshop on Software Technolgies for Embedded and Ubiquitous Systems*. Springer. 2010, pp. 35–46.

Chapter 8
Conclusions and Future Work

In this chapter, first, the conclusion from the book is presented and then some open issues for future works are discussed.

8.1 Conclusions

MC systems are getting more attention in the last decade due to their significance in safety-critical applications, such as medical devices like artificial heart, and avionics like flight control, etc. These MC systems have been devised to address the real-time and safety requirements of these industrial safety-critical applications, where applications with different criticality levels are integrated into a common hardware platform to reduce cost, size, and power consumption. The main research question is how to reconcile the requirements of reliability and real time while improving resource utilization. This question raises problems in modeling, designing at application and software levels, and implementing and controlling the hardware.

The design of such MC system comes with certain challenges, which are mainly faced in the MC application analysis, how the tasks' parameters are defined, scheduling analysis, and MC hardware analysis. In application-level analysis, due to defining multiple WCETs for each task, corresponding to the multiple criticality levels and the ongoing mode of operation, the timing behavior of MC systems (which correlates with the system mode switching probability and utilization) would not be stable at run-time. Determining the appropriate values of WCETs for lower criticality modes is nontrivial and needs to be addressed in MC system design. Besides, from MC task scheduling analysis perspective, the existing MC task scheduling algorithms, like EDF-VD, drop/degrade the LC tasks in the higher modes. However, the frequent drop of LC tasks, such as mission-critical tasks in some safety-critical applications, may have a negative impact on the execution of other tasks, like HC tasks and mission-critical tasks themselves.

B. Ranjbar et al., *Quality-of-Service Aware Design and Management of Embedded Mixed-Criticality Systems*, https://doi.org/10.1007/978-3-031-38960-3_8

Although it does not cause catastrophic consequences, it may prevent the system from accomplishing its mission correctly. To this end, the number of allowable drops for each LC task must be restricted in such MC systems and should be studied. From the perspective of the MC hardware system design, the critical trend of MC system design is integrating functions with different criticality levels onto a common hardware platform, where the platforms can be single- or multi-core architecture. One of the goals in designing the MC systems is resource utilization improvement. However, the execution of LC and HC tasks require higher computational demands, which leads to high-power consumption. Systems with high power are more likely to generate unexpected heat beyond the cooling capacity. They will be more susceptible to failures and instability, which is unacceptable for MC systems and may cause catastrophic consequences. This heat generation would be harmful if the degree of freedom (regarding the availability of the cores) increases due to heat transmission among cores. Although employing multi-core platforms helps to improve the QoS by executing the tasks in parallel, guaranteeing the real-time constraints while managing the system power consumption is a crucial challenge that must be addressed.

In this book, we addressed the mentioned challenges in MC application and hardware system designs to improve the QoS while guaranteeing the real-time constraints of tasks. In Chap. 3, since the MC system parameters like WCET are the key aspects of system design in application-level analysis, we focused on WCET estimation of MC tasks in the LO mode. In this chapter, an analytical scheme, called *BOT-MICS*, was first proposed to determine the WCET at design-time, to make a trade-off between the number of scheduled tasks at design-time (i.e., utilization) and the number of dropped LC tasks at run-time as a result of frequent mode switches. The analytical scheme is based on the *Chebyshev theorem* which shows the relation between the low WCETs and mode switching probability. We formulated the problem and solved it using GA to improve the objectives, resource utilization, and mode switching probability. We evaluated the *BOT-MICS* for various state-of-the-art MC systems, and therefore, the experimental results showed that it improves the utilization of state-of-the-art MC systems by up to 85.29% while maintaining 9.11% mode switching probability in the worst-case scenario. However, although *BOT-MICS* can determine the optimum values of low WCETs at design-time, these values are static and remain unchanged for each task, which cannot adapt to task dynamism at run-time. It can cause processor underutilization if the low WCETs are not close to the actual execution times. Therefore, we proposed *ADAPTIVE* to determine the low WCET at run-time based on the actual execution times of tasks. We considered the run-time behavior of tasks and proposed a learning-based approach that dynamically monitors the tasks' execution times and adapts the low WCETs to improve QoS at the end of complete application execution. Based on our observations on running embedded real-time benchmarks on a real platform, *ADAPTIVE* can improve the QoS by 16.4% on average while reducing the utilization waste by 17.7% on average, compared to state-of-the-art works.

Further, we focused on MC task scheduling analysis and proposed methods in Chaps. 4 and 5, to schedule more LC tasks in the HI mode and improve the QoS.

To this end, we proposed *FANTOM* in Chap. 4, in which we first introduced a task parameter to limit the number of LC task drops. Then, we developed a design-time task-drop-aware schedulability analysis based on the EDF-VD in accordance with the defined parameter. By defining the new parameter, we consider a maximum allowable number of drops for each LC task and prohibit the number of drops from passing a predefined threshold. According to the obtained results from an extensive set of simulations, which have been validated through a realistic avionic application task set, *FANTOM* improves the acceptance ratio by up to 43.9% compared to a state-of-the-art work. Since *FANTOM* is proposed based on the worst-case scenario of task execution times, we then proposed a learning-based drop-aware task scheduling mechanism in Chap. 5. Since the tasks are not executed always up to their WCET, the proposed approach carefully monitors the alterations in the behavior of the MC system at run-time, to exploit the generated dynamic slack for reducing the LC tasks' penalty and preventing frequent drops of LC tasks in the future. Based on an extensive set of experiments, our observations have shown that the proposed approach exploits accumulated dynamic slack generated at run-time, by 9.84% more on average compared to existing works, and can reduce the deadline miss rate by up to 51.78% and 33.27% on average, compared to state-of-the-art works.

By increasing the number of tasks in an application, although multi-core platforms are fruitful due to the ability of parallel execution of tasks on cores, the power consumption of these systems is a critical issue. Therefore, we studied the challenges of these MC hardware design in Chaps. 6 and 7. First, we proposed an approach in fault-tolerant multi-core MC systems to manage peak power consumption and temperature. The approach develops a tree of possible task mapping and scheduling at design-time to cover all possible scenarios and reduce the LC task drop rate in the HI mode, i.e., improve the QoS. At run-time, the system exploits the tree to select a proper schedule according to fault occurrences (to guarantee the real-time constraints in case of fault occurrence) and criticality mode changes. Experimental results show that the average task schedulability is 74.14% on average while improving the peak power consumption and maximum temperature by 16.65% and 14.9 °C on average, respectively, compared to recent work. In addition, for a real-life application, our method reduces the peak power and maximum temperature by up to 20.06% and 5 °C, respectively, compared to a state-of-the-art work. In order to take advantage of task dynamism in run-time system operation, we then proposed an online peak power and thermal management heuristic in Chap. 7, in which the re-mapping technique is used in the case of available dynamic slack to re-map a ready task from the hot core to a core with a lower temperature to manage the system's maximum temperature. This heuristic exploits the generated dynamic slack (due to early completion of a task execution) and assigns them to an appropriate task among k look-ahead tasks, which has more impact on system power and maximum temperature and reduces the *V-f* levels. However, changing the frequency and selecting a proper task for slack assignment and a proper core for task re-mapping at run-time can be time-consuming and may cause deadline violation which is not admissible for HC tasks. Therefore, we analyzed and then optimized the proposed run-time scheduler and evaluate it for various platforms. The proposed approach was experimentally validated on the ODROID-XU3 with various embedded real-time benchmarks. Results show that the heuristic achieves up to 5.25% reduction in system peak power and 20.33% (16.8 °C) reduction in maximum temperature compared to an existing method while meeting deadline constraints in different criticality modes.

8.2 Future Work

In this book, system-level approaches are proposed through MC application analysis and MC hardware analysis to enhance the QoS of MC systems while ensuring real-time constraints. There are, however, several open issues that need to be addressed when designing such MC systems, including the following:

- **MC System Analysis with Consideration of Communication and Data Sharing**: Although employing multi-core platforms offers the opportunity for executing several applications in common hardware, safety and real-timeliness are critical issues in designing such MC hardware systems. In these multi-core platforms, more data are needed to be shared between concurrent executions of tasks with different criticality levels [1]. The strict control of data (critical and noncritical), communication, sharing, and storage in such systems for safety assurance, e.g., in medical devices, is crucial. It is essential to ensure that the behavior of LC tasks does not adversely affect the behavior of HC tasks while communicating and sharing data or writing in memories, for example, delaying an HC task by blocking the memory access. Most state-of-the-art works have concentrated on designing MC systems in multi-core processors regardless of safe and on-time data sharing among communication and memories. As a result, an MC system design considering all system resources, like communications, memory access, and processors, needs to be developed.
- **Evaluation on Real Platforms Using Real-Life MC Benchmarks**: Most approaches in the field of MC systems are presented in the academic area, not evaluated by industries, and may not be usable in reality. For this purpose of proposing more practical approaches, first, the designers need more realistic MC task and system models that are derived based on industrial application features and run-time behavior, which is good to be focused on as future work. Besides, some open-access real-life MC benchmarks are also fruitful and need to be studied and presented, e.g., in automotive and avionic industrial applications, for analyzing and evaluating presented approaches in MC system design domain. Then, since most proposed methods in the academic area are evaluated through simulation, an evaluation and implementation of approaches on a real embedded multi-core platform are needed to demonstrate their effectiveness. It is useful if the chosen platforms are mostly the same as those used in industries, like automotive platforms.

- **Safety and Reliability Management**: In MC systems, the correct execution of functions, especially HC tasks, must be ensured during run-time under various stresses (e.g., hardware errors, software errors, etc.) to prevent failure and catastrophic consequences. Therefore, MC systems must be well designed to ensure long-term and application-specific reliability. In order to guarantee system safety, fault-tolerance techniques are employed in the design of such systems. In the case of fault occurrence, different techniques such as replication or re-execution are needed to enhance their strengths against potential failures. Furthermore, due to the various safety demands for tasks, they can have different reliability requirements. A failure occurring in tasks with different criticality levels has a disparate impact on the system, from no effect to catastrophic. Although there are some approaches that have focused on managing the lifetime or timing reliability in hardware or system software layers, in future work, more works are first needed to analyze the MC systems' behavior in all abstraction layers like hardware, application, and system software, separately, while managing the reliability. These single-layer reliability-aware design approaches adopt an other-layer-agnostic approach [2]. However, this isolated layer-wise fault mitigation has a high cost (in terms of power, area, and timing), making it infeasible for most embedded systems [3]. Therefore, considering the cross-layer reliability is also crucial in efficiently designing these MC systems that need to be focused on and deeply studied in the future.
- **Ensuring ML-Based Objective (QoS, Reliability), with Consideration of ML Complexity Reduction**: In general, there are three categories of ML techniques—supervised learning, unsupervised learning, and reinforcement learning—where, depending on the problem, parameters, and inputs, at least one of these techniques is determined and used for system property improvement. These ML techniques are usually memory-intensive and computationally expensive, which makes some of them incompatible with embedded MC system design. Further, a more simplified ML technique may not provide the desired accuracy when making predictions. Therefore, although we have proposed ML-based approaches for QoS-aware MC system design, these ML techniques are needed to be investigated first in terms of overheads, accuracy, and capability, and then the efficient ones which are suitable for a specified objective to be applied to embedded MC systems at design-time are also needed to be determined. Besides, to use the ML techniques at run-time, the selected ML techniques must be light enough with low overhead to be used at run-time for objective improvement in MC systems, which are safety-critical, with no effect on MC applications' timeliness. The ML technique's function is to choose the best decision under environmental changes in order to improve objectives, like reliability, and QoS. Therefore, selecting a suitable ML technique, concentrating on its run-time timing overheads and accuracy for learning and prediction, plays a crucial role in controlling and doing a safe mission under environmental changes and needs to be intensely focused. Besides, although using ML techniques can improve the objectives, they cannot guarantee the constraints and requirements. For example, some research works have been proposed to enhance reliability through ML

techniques, but there is no guarantee of meeting the reliability requirements. Designing and developing ML-based approaches to guarantee the reliability requirements of MC tasks would be interesting to be noticed.

- **Hardware- and Overhead-Aware WCET Estimation**: This book focused on how the low WCET of MC tasks are determined while improving their utilization. In the proposed approaches, a single-core platform has been considered in designing the MC systems with no memory or communication consideration. This is despite the fact that some other factors impact WCET values and processor utilization, like hardware behavior, overheads, and task dependencies, and make the system inaccurately designed. In the following, we discuss briefly how these factors may have a negative impact on designing MC systems.
  - Overheads and task dependencies: In general, the WCET values depend on the function's input data. This fact would be more important if the tasks are dependent. Tasks should be analyzed at the instruction level to see how the variety of input data can affect their WCETs. Besides, some timing overheads like saving the output result of task execution in memory/cache and loading from to be used as a function input can be significant in real-time systems. It is important to consider how overheads arising from tasks with one criticality level may affect tasks with a different criticality level. As a result, it is necessary to find first appropriate models of system overheads and task dependencies and then integrate them into the MC's system analysis: the WCET estimation and utilization improvement.
  - Hardware behavior analysis: In order to estimate low WCET of MC tasks, in this book, we have analyzed the tasks, predicted the low WCETs, and observed on a concrete processor. However, as mentioned in the first future work, a hardware platform consists of the computation part, communication, and memory. These parts affect the execution times of MC tasks and make them context-dependent. To estimate the appropriate low WCET of tasks to be used in MC system design, all hardware components' behavior should be analyzed for all paths leading to a task's instructions. Considering the hardware behavior analysis in application-level analysis leads to better MC system design, which needs to be focused on.

References

1. Alan Burns and Robert I. Davis. "A Survey of Research into Mixed-Criticality Systems". In: *ACM Computing Surveys (CSUR)*, 50.6 (2017), pp. 1–37.
2. Siva Satyendra Sahoo, Bharadwaj Veeravalli, and A. Kumar. "A Hybrid Agent-Based Design Methodology for Dynamic Cross-Layer Reliability in Heterogeneous Embedded Systems". In: *Proc. of ACM/IEEE Design Automation Conference (DAC)*, 2019.
3. Siva Satyendra Sahoo, Bharadwaj Veeravalli, and A. Kumar. "Cross-layer fault-tolerant design of real-time systems". In: *Proc. of IEEE International Symposium on Defect and Fault Tolerance in VLSI and Nanotechnology Systems (DFT)*, 2016, pp. 63–68.

Index

B. Ranjbar et al., *Quality-of-Service Aware Design and Management of Embedded Mixed-Criticality Systems*, https://doi.org/10.1007/978-3-031-38960-3

The manufacturer's authorised representative in the EU is Springer Nature Customer Service Centre GmbH, Europaplatz 3, 69115 Heidelberg, Germany. If you have any concerns regarding our products, please contact ProductSafety@springernature.com

Printed and bound by CPI Group (UK) Ltd, Croydon, CR0 4YY

01/08/2026

02183512-0001